Ben Stacy Jerrik (Ed.)

# Semi-Symmetric Graph

Ben Stacy Jerrik (Ed.)

# Semi-Symmetric Graph

## Semi-Symmetric Graph

Part Press

**Imprint**

Permission is granted to copy, distribute and/or modify this document under the terms of the GNU Free Documentation License, Version 1.2 or any later version published by the Free Software Foundation; with no Invariant Sections, with the Front-Cover Texts, and with the Back- Cover Texts. A copy of the license is included in the section entitled "GNU Free Documentation License".

All parts of this book are extracted from Wikipedia, the free encyclopedia (www.wikipedia.org).

You can get detailed informations about the authors of this collection of articles at the end of this book. The editors (Ed.) of this book are no authors. They have not modified or extended the original texts.

Pictures published in this book can be under different licences than the GNU Free Documentation License. You can get detailed informations about the authors and licences of pictures at the end of this book.

The content of this book was generated collaboratively by volunteers. Please be advised that nothing found here has necessarily been reviewed by people with the expertise required to provide you with complete, accurate or reliable information. Some information in this book maybe misleading or wrong. The Publisher does not guarantee the validity of the information found here. If you need specific advice (f.e. in fields of medical, legal, financial, or risk management questions) please contact a professional who is licensed or knowledgeable in that area.

Any brand names and product names mentioned in this book are subject to trademark, brand or patent protection and are trademarks or registered trademarks of their respective holders. The use of brand names, product names, common names, trade names, product descriptions etc. even without a particular marking in this works is in no way to be construed to mean that such names may be regarded as unrestricted in respect of trademark and brand protection legislation and could thus be used by anyone.

Cover image: www.ingimage.com
Concerning the licence of the cover image please contact ingimage.

Publisher:
Part Press is a trademark of
International Book Market Service Ltd., 17 Rue Meldrum, Beau Bassin, 1713-01 Mauritius
Email: info@bookmarketservice.com
Website: www.bookmarketservice.com

Published in 2011

Printed in: U.S.A., U.K., Germany. This book was not produced in Mauritius.

**ISBN: 978-613-8-75478-7**

# Contents

## Articles



## References

# Semi-symmetric graph

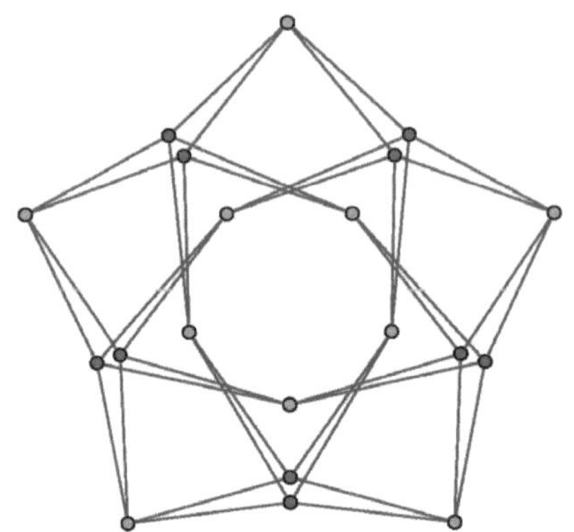

The Folkman graph, the smallest semi-symmetric graph.

| Graph families defined by their automorphisms | | |
|---|---|---|
| distance-transitive | distance-regular | strongly regular |
| symmetric (arc-transitive) | $t$-transitive, $t \geq 2$ | |
| (if connected) | | |
| vertex- and edge-transitive | edge-transitive and regular | edge-transitive |
| vertex-transitive | regular | |
| Cayley graph | skew-symmetric | asymmetric |

In the mathematical field of graph theory, a **semi-symmetric graph** is an undirected graph that is edge-transitive and regular, but not vertex-transitive.

In other words, a graph is semi-symmetric if each vertex has the same number of incident edges, and there is a symmetry taking any of its edges to any other of its edges, but there is some pair of vertices that cannot be mapped into each other by a symmetry. A semi-symmetric graph must be bipartite, and its automorphism group must act transitively on each of the two vertex sets of the bipartition. In the diagram on the right, green vertices can not be mapped to red ones by any automorphism.

Semi-symmetric graphs were first studied by Jon Folkman in 1967, who discovered the smallest semi-symmetric graph, the Folkman graph on 20 vertices.[1]

The smallest cubic semi-symmetric graph is the Gray graph on 54 vertices. It was first observed to be semi-symmetric by Bouwer (1968). It was proven to be the smallest cubic semi-symmetric graph by Dragan Marušič and Aleksander Malnič.[2]

All the cubic semi-symmetric graphs on up to 768 vertices are known. According to Conder, Malnič, Marušič and Potočnik, the four smallest possible cubic semi-symmetric graph after the Gray graph are the Iofinova–Ivanov graph on 110 vertices, the Ljubljana graph on 112 vertices,[3] a graph on 120 vertices with girth 8 and the Tutte 12-cage.[4]

## References

[1]  Folkman, J. (1967), "Regular line-symmetric graphs", *Journal of Combinatorial Theory* **3** (3): 215–232, doi:10.1016/S0021-9800(67)80069-3.

[2]  Bouwer, I. Z. (1968), "An edge but not vertex transitive cubic graph", *Bulletin of the Canadian Mathematical Society* **11**: 533–535, doi:10.4153/CMB-1968-063-0.

[3]  Conder, M.; Malnič, A.; Marušič, D.; Pisanski, T.; Potočnik, P. (2002), "The Ljubljana Graph" (http://www.imfm.si/preprinti/PDF/ 00845.pdf), *IMFM Preprints* (Ljubljana: Institute of Mathematics, Physics and Mechanics) **40** (845), .

[4]  Conder, Marston; Malnič, Aleksander; Marušič, Dragan; Potočnik, Primož (2006), "A census of semisymmetric cubic graphs on up to 768 vertices", *Journal of Algebraic Combinatorics* **23** (3): 255–294, doi:10.1007/s10801-006-7397-3.

## External links

- Weisstein, Eric W., " Semisymmetric Graph (http://mathworld.wolfram.com/SemisymmetricGraph.html)" from MathWorld.

# Graph theory

In mathematics and computer science, **graph theory** is the study of *graphs*, mathematical structures used to model pairwise relations between objects from a certain collection. A "graph" in this context refers to a collection of vertices or 'nodes' and a collection of *edges* that connect pairs of vertices. A graph may be *undirected*, meaning that there is no distinction between the two vertices associated with each edge, or its edges may be *directed* from one vertex to another; see graph (mathematics) for more detailed definitions and for other variations in the types of graphs that are commonly considered. The graphs studied in graph theory should not be confused with graphs of functions or other kinds of graphs.

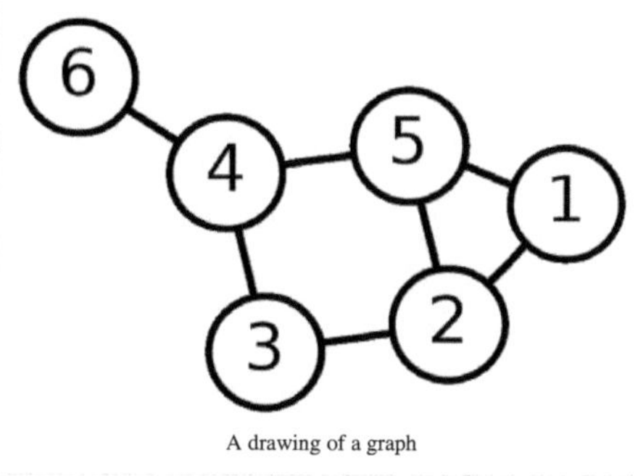
A drawing of a graph

Graphs are one of the prime objects of study in discrete mathematics. Refer to glossary of graph theory for basic definitions in graph theory.

## Applications

Graphs are among the most ubiquitous models of both natural and human-made structures. They can be used to model many types of relations and process dynamics in physical, biological[1] and social systems. Many problems of practical interest can be represented by graphs.

In computer science, graphs are used to represent networks of communication, data organization, computational devices, the flow of computation, etc. One practical example: The link structure of a website could be represented by a directed graph. The vertices are the web pages available at the website and a directed edge from page *A* to page *B* exists if and only if *A* contains a link to *B*. A similar approach can be taken to problems in travel, biology, computer chip design, and many other fields. The development of algorithms to handle graphs is therefore of major interest in computer science. There, the transformation of graphs is often formalized and represented by graph rewrite systems. They are either directly used or properties of the rewrite systems (e.g. confluence) are studied. Complementary to graph transformation systems focussing on rule-based in-memory manipulation of graphs are graph databases geared towards transaction-safe, persistent storing and querying of graph-structured data.

Graph-theoretic methods, in various forms, have proven particularly useful in linguistics, since natural language often lends itself well to discrete structure. Traditionally, syntax and compositional semantics follow tree-based structures, whose expressive power lies in the Principle of Compositionality, modeled in a hierarchical graph. Within lexical semantics, especially as applied to computers, modeling word meaning is easier when a given word is understood in terms of related words; semantic networks are therefore important in computational linguistics. Still other methods in phonology (e.g. Optimality Theory, which uses lattice graphs) and morphology (e.g. finite-state morphology, using finite-state transducers) are common in the analysis of language as a graph. Indeed, the usefulness of this area of mathematics to linguistics has borne organizations such as TextGraphs [2], as well as various 'Net' projects, such as WordNet, VerbNet, and others.

Graph theory is also used to study molecules in chemistry and physics. In condensed matter physics, the three dimensional structure of complicated simulated atomic structures can be studied quantitatively by gathering statistics on graph-theoretic properties related to the topology of the atoms. For example, Franzblau's shortest-path (SP) rings. In chemistry a graph makes a natural model for a molecule, where vertices represent atoms and edges bonds. This approach is especially used in computer processing of molecular structures, ranging from chemical editors to database searching. In statistical physics, graphs can represent local connections between interacting parts of a system, as well as the dynamics of a physical process on such systems.

Graph theory is also widely used in sociology as a way, for example, to measure actors' prestige or to explore diffusion mechanisms, notably through the use of social network analysis software.

Likewise, graph theory is useful in biology and conservation efforts where a vertex can represent regions where certain species exist (or habitats) and the edges represent migration paths, or movement between the regions. This information is important when looking at breeding patterns or tracking the spread of disease, parasites or how changes to the movement can affect other species.

In mathematics, graphs are useful in geometry and certain parts of topology, e.g. Knot Theory. Algebraic graph theory has close links with group theory.

A graph structure can be extended by assigning a weight to each edge of the graph. Graphs with weights, or weighted graphs, are used to represent structures in which pairwise connections have some numerical values. For example if a graph represents a road network, the weights could represent the length of each road.

A digraph with weighted edges in the context of graph theory is called a network. Network analysis have many practical applications, for example, to model and analyze traffic networks. Applications of network analysis split broadly into three categories:

1. First, analysis to determine structural properties of a network, such as the distribution of vertex degrees and the diameter of the graph. A vast number of graph measures exist, and the production of useful ones for various domains remains an active area of research.
2. Second, analysis to find a measurable quantity within the network, for example, for a transportation network, the level of vehicular flow within any portion of it.
3. Third, analysis of dynamical properties of networks.

## History

The first text book on graph theory was written by Denes Konig, but this book was very hard to understand. Richard Rado said he was unable to understand Konig's proof of Rado's Theorem. The first readable book was by Frank Harary and was enormously popular. This book enabled mathematicians, chemists, electrical engineers and social scientists to talk to each other. He donated all of the proceeds to fund the Polya Prize.

The paper written by Leonhard Euler on the *Seven Bridges of Königsberg* and published in 1736 is regarded as the first paper in the history of graph theory.[3] This paper, as well as the one written by Vandermonde on the *knight problem,* carried on with the *analysis situs* initiated by Leibniz. Euler's formula relating the number of edges, vertices, and faces of a convex polyhedron was studied and generalized by Cauchy[4] and L'Huillier,[5] and is at the origin of topology.

More than one century after Euler's paper on the bridges of Königsberg and while Listing introduced topology, Cayley was led by the study of particular analytical forms arising from differential calculus to study a particular class of graphs, the *trees.* This study

The Königsberg Bridge problem

had many implications in theoretical chemistry. The involved techniques mainly concerned the enumeration of graphs having particular properties. Enumerative graph theory then rose from the results of Cayley and the fundamental results published by Pólya between 1935 and 1937 and the generalization of these by De Bruijn in 1959. Cayley linked his results on trees with the contemporary studies of chemical composition.[6] The fusion of the ideas coming from mathematics with those coming from chemistry is at the origin of a part of the standard terminology of graph theory.

In particular, the term "graph" was introduced by Sylvester in a paper published in 1878 in *Nature*, where he draws an analogy between "quantic invariants" and "co-variants" of algebra and molecular diagrams:[7]

> "[...] Every invariant and co-variant thus becomes expressible by a *graph* precisely identical with a Kekuléan diagram or chemicograph. [...] I give a rule for the geometrical multiplication of graphs, *i.e.* for constructing a *graph* to the product of in- or co-variants whose separate graphs are given. [...]" (italics as in the original).

One of the most famous and productive problems of graph theory is the four color problem: "Is it true that any map drawn in the plane may have its regions colored with four colors, in such a way that any two regions having a common border have different colors?" This problem was first posed by Francis Guthrie in 1852 and its first written record is in a letter of De Morgan addressed to Hamilton the same year. Many incorrect proofs have been proposed, including those by Cayley, Kempe, and others. The study and the generalization of this problem by Tait, Heawood, Ramsey and Hadwiger led to the study of the colorings of the graphs embedded on surfaces with arbitrary genus. Tait's reformulation generated a new class of problems, the *factorization problems*, particularly studied by Petersen and Kőnig. The works of Ramsey on colorations and more specially the results obtained by Turán in 1941 was at the origin of another branch of graph theory, *extremal graph theory.*

The four color problem remained unsolved for more than a century. In 1969 Heinrich Heesch published a method for solving the problem using computers.[8] A computer-aided proof produced in 1976 by Kenneth Appel and Wolfgang Haken makes fundamental use of the notion of "discharging" developed by Heesch.[9] [10] The proof involved checking the properties of 1,936 configurations by computer, and was not fully accepted at the time due to its complexity. A simpler proof considering only 633 configurations was given twenty years later by Robertson, Seymour, Sanders and Thomas.[11]

The autonomous development of topology from 1860 and 1930 fertilized graph theory back through the works of Jordan, Kuratowski and Whitney. Another important factor of common development of graph theory and topology came from the use of the techniques of modern algebra. The first example of such a use comes from the work of the physicist Gustav Kirchhoff, who published in 1845 his Kirchhoff's circuit laws for calculating the voltage and current in electric circuits.

The introduction of probabilistic methods in graph theory, especially in the study of Erdős and Rényi of the asymptotic probability of graph connectivity, gave rise to yet another branch, known as *random graph theory*, which has been a fruitful source of graph-theoretic results.

# Drawing graphs

Graphs are represented graphically by drawing a dot or circle for every vertex, and drawing an arc between two vertices if they are connected by an edge. If the graph is directed, the direction is indicated by drawing an arrow.

A graph drawing should not be confused with the graph itself (the abstract, non-visual structure) as there are several ways to structure the graph drawing. All that matters is which vertices are connected to which others by how many edges and not the exact layout. In practice it is often difficult to decide if two drawings represent the same graph. Depending on the problem domain some layouts may be better suited and easier to understand than others.

# Graph-theoretic data structures

There are different ways to store graphs in a computer system. The data structure used depends on both the graph structure and the algorithm used for manipulating the graph. Theoretically one can distinguish between list and matrix structures but in concrete applications the best structure is often a combination of both. List structures are often preferred for sparse graphs as they have smaller memory requirements. Matrix structures on the other hand provide faster access for some applications but can consume huge amounts of memory.

## List structures

Incidence list

> The edges are represented by an array containing pairs (tuples if directed) of vertices (that the edge connects) and possibly weight and other data. Vertices connected by an edge are said to be *adjacent*.

Adjacency list

> Much like the incidence list, each vertex has a list of which vertices it is adjacent to. This causes redundancy in an undirected graph: for example, if vertices A and B are adjacent, A's adjacency list contains B, while B's list contains A. Adjacency queries are faster, at the cost of extra storage space.

## Matrix structures

Incidence matrix

> The graph is represented by a matrix of size $|V|$ (number of vertices) by $|E|$ (number of edges) where the entry [vertex, edge] contains the edge's endpoint data (simplest case: 1 - incident, 0 - not incident).

Adjacency matrix

> This is an $n$ by $n$ matrix $A$, where $n$ is the number of vertices in the graph. If there is an edge from a vertex $x$ to a vertex $y$, then the element is 1 (or in general the number of $xy$ edges), otherwise it is 0. In computing, this matrix makes it easy to find subgraphs, and to reverse a directed graph.

Laplacian matrix or "Kirchhoff matrix" or "Admittance matrix"

> This is defined as $D - A$, where $D$ is the diagonal degree matrix. It explicitly contains both adjacency information and degree information. (However, there are other, similar matrices that are also called "Laplacian matrices" of a graph.)

Distance matrix

> A symmetric $n$ by $n$ matrix $D$, where $n$ is the number of vertices in the graph. The element is the length of a shortest path between $x$ and $y$; if there is no such path = infinity. It can be derived from powers of $A$

# Problems in graph theory

## Enumeration

There is a large literature on graphical enumeration: the problem of counting graphs meeting specified conditions. Some of this work is found in Harary and Palmer (1973).

## Subgraphs, induced subgraphs, and minors

A common problem, called the subgraph isomorphism problem, is finding a fixed graph as a subgraph in a given graph. One reason to be interested in such a question is that many graph properties are *hereditary* for subgraphs, which means that a graph has the property if and only if all subgraphs have it too. Unfortunately, finding maximal subgraphs of a certain kind is often an NP-complete problem.

- Finding the largest complete graph is called the clique problem (NP-complete).

A similar problem is finding induced subgraphs in a given graph. Again, some important graph properties are hereditary with respect to induced subgraphs, which means that a graph has a property if and only if all induced subgraphs also have it. Finding maximal induced subgraphs of a certain kind is also often NP-complete. For example,

- Finding the largest edgeless induced subgraph, or independent set, called the independent set problem (NP-complete).

Still another such problem, the *minor containment problem*, is to find a fixed graph as a minor of a given graph. A minor or **subcontraction** of a graph is any graph obtained by taking a subgraph and contracting some (or no) edges. Many graph properties are hereditary for minors, which means that a graph has a property if and only if all minors have it too. A famous example:

- A graph is planar if it contains as a minor neither the complete bipartite graph (See the Three-cottage problem) nor the complete graph .

Another class of problems has to do with the extent to which various species and generalizations of graphs are determined by their *point-deleted subgraphs*, for example:

- The reconstruction conjecture.

## Graph coloring

Many problems have to do with various ways of coloring graphs, for example:

- The four-color theorem
- The strong perfect graph theorem
- The Erdős–Faber–Lovász conjecture (unsolved)
- The total coloring conjecture (unsolved)
- The list coloring conjecture (unsolved)
- The Hadwiger conjecture (graph theory) (unsolved).

## Route problems

- Hamiltonian path and cycle problems
- Minimum spanning tree
- Route inspection problem (also called the "Chinese Postman Problem")
- Seven Bridges of Königsberg
- Shortest path problem
- Steiner tree
- Three-cottage problem
- Traveling salesman problem (NP-hard)

## Network flow

There are numerous problems arising especially from applications that have to do with various notions of flows in networks, for example:

- Max flow min cut theorem

## Visibility graph problems

- Museum guard problem

## Covering problems

Covering problems are specific instances of subgraph-finding problems, and they tend to be closely related to the clique problem or the independent set problem.

- Set cover problem
- Vertex cover problem

## Graph classes

Many problems involve characterizing the members of various classes of graphs. Overlapping significantly with other types in this list, this type of problem includes, for instance:

- Enumerating the members of a class
- Characterizing a class in terms of forbidden substructures
- Ascertaining relationships among classes (e.g., does one property of graphs imply another)
- Finding efficient algorithms to decide membership in a class
- Finding representations for members of a class.

# See also

- Gallery of named graphs
- Glossary of graph theory
- List of graph theory topics
- Publications in graph theory

## Related topics

- Graph property
- Algebraic graph theory
- Conceptual graph
- Data structure

- Disjoint-set data structure
- Entitative graph
- Existential graph
- Graph data structure
- Graph algebras
- Graph automorphism
- Graph coloring
- Graph database
- Graph drawing
- Graph equation
- Graph rewriting
- Graph sandwich problem
- Intersection graph
- Logical graph
- Loop
- Network theory
- Null graph
- Pebble motion problems
- Percolation
- Perfect graph
- Quantum graph
- Random regular graphs
- Spectral graph theory
- Strongly regular graphs
- Symmetric graphs
- Tree data structure

## Algorithms

- Bellman-Ford algorithm
- Dijkstra's algorithm
- Ford-Fulkerson algorithm
- Kruskal's algorithm
- Nearest neighbour algorithm
- Prim's algorithm
- Depth-first search
- Breadth-first search

## Subareas

- Algebraic graph theory
- Geometric graph theory
- Extremal graph theory
- Probabilistic graph theory
- Topological graph theory

### Related areas of mathematics

- Combinatorics
- Group theory
- Knot theory
- Ramsey theory

### Generalizations

- Hypergraph
- Abstract simplicial complex

### Prominent graph theorists

- Alon, Noga
- Berge, Claude
- Bollobás, Béla
- Brightwell, Graham
- Chung, Fan
- Dirac, Gabriel Andrew
- Erdős, Paul
- Euler, Leonhard
- Faudree, Ralph
- Golumbic, Martin
- Graham, Ronald
- Harary, Frank
- Heawood, Percy John
- Kőnig, Dénes
- Lovász, László
- Nešetřil, Jaroslav
- Rényi, Alfréd
- Ringel, Gerhard
- Robertson, Neil
- Seymour, Paul
- Szemerédi, Endre
- Thomas, Robin
- Thomassen, Carsten
- Turán, Pál
- Tutte, W. T.
- Whitney, Hassler

## Notes

[1]  Mashaghi, A.; *et al.* (2004). "Investigation of a protein complex network". *European Physical Journal B* **41** (1): 113–121. doi:10.1140/epjb/e2004-00301-0.

[2]  http://www.textgraphs.org

[3]  Biggs, N.; Lloyd, E. and Wilson, R. (1986), *Graph Theory, 1736-1936*, Oxford University Press

[4]  Cauchy, A.L. (1813), "Recherche sur les polyèdres - premier mémoire", *Journal de l'Ecole Polytechnique* **9 (Cahier 16)**: 66–86.

[5]  L'Huillier, S.-A.-J. (1861), "Mémoire sur la polyèdrométrie", *Annales de Mathématiques* **3**: 169–189.

[6]  Cayley, A. (1875), "Ueber die Analytischen Figuren, welche in der Mathematik Bäume genannt werden und ihre Anwendung auf die Theorie chemischer Verbindungen", *Berichte der deutschen Chemischen Gesellschaft* **8** (2): 1056–1059, doi:10.1002/cber.18750080252.

[7]  John Joseph Sylvester (1878), *Chemistry and Algebra*. Nature, volume 17, page 284. doi:10.1038/017284a0. Online version (http://www. archive.org/stream/nature15unkngoog#page/n312/mode/1up). Retrieved 2009-12-30.

[8]  Heinrich Heesch: Untersuchungen zum Vierfarbenproblem. Mannheim: Bibliographisches Institut 1969.

[9]  Appel, K. and Haken, W. (1977), "Every planar map is four colorable. Part I. Discharging", *Illinois J. Math.* **21**: 429–490.

[10]  Appel, K. and Haken, W. (1977), "Every planar map is four colorable. Part II. Reducibility", *Illinois J. Math.* **21**: 491–567.

[11]  Robertson, N.; Sanders, D.; Seymour, P. and Thomas, R. (1997), "The four color theorem", *Journal of Combinatorial Theory Series B* **70**: 2–44, doi:10.1006/jctb.1997.1750.

## References

- Berge, Claude (1958), *Théorie des graphes et ses applications*, Collection Universitaire de Mathématiques, **II**, Paris: Dunod. English edition, Wiley 1961; Methuen & Co, New York 1962; Russian, Moscow 1961; Spanish, Mexico 1962; Roumanian, Bucharest 1969; Chinese, Shanghai 1963; Second printing of the 1962 first English edition, Dover, New York 2001.
- Biggs, N.; Lloyd, E.; Wilson, R. (1986), *Graph Theory, 1736–1936*, Oxford University Press.
- Bondy, J.A.; Murty, U.S.R. (2008), *Graph Theory*, Springer, ISBN 978-1-84628-969-9.
- Bondy, Riordan, O.M (2003), *Mathematical results on scale-free random graphs in "Handbook of Graphs and Networks" (S. Bornholdt and H.G. Schuster (eds)), Wiley VCH, Weinheim, 1st ed.*.
- Chartrand, Gary (1985), *Introductory Graph Theory*, Dover, ISBN 0-486-24775-9.
- Gibbons, Alan (1985), *Algorithmic Graph Theory*, Cambridge University Press.
- Reuven Cohen, Shlomo Havlin (2010), *Complex Networks: Structure, Robustness and Function*, Cambridge University Press
- Golumbic, Martin (1980), *Algorithmic Graph Theory and Perfect Graphs*, Academic Press.
- Harary, Frank (1969), *Graph Theory*, Reading, MA: Addison-Wesley.
- Harary, Frank; Palmer, Edgar M. (1973), *Graphical Enumeration*, New York, NY: Academic Press.
- Mahadev, N.V.R.; Peled, Uri N. (1995), *Threshold Graphs and Related Topics*, North-Holland.
- Mark Newman (2010), *Networks: An Introduction*, Oxford University Press.

## External links

### Online textbooks

- Graph Theory with Applications (http://www.math.jussieu.fr/~jabondy/books/gtwa/gtwa.html) (1976) by Bondy and Murty
- Phase Transitions in Combinatorial Optimization Problems, Section 3: Introduction to Graphs (http://arxiv.org/pdf/cond-mat/0602129) (2006) by Hartmann and Weigt
- Digraphs: Theory Algorithms and Applications (http://www.cs.rhul.ac.uk/books/dbook/) 2007 by Jorgen Bang-Jensen and Gregory Gutin
- Graph Theory, by Reinhard Diestel (http://diestel-graph-theory.com/index.html)

## Other resources

- Graph theory tutorial (http://www.utm.edu/departments/math/graph/)
- A searchable database of small connected graphs (http://www.gfredericks.com/main/sandbox/graphs)
- Image gallery: graphs (http://web.archive.org/web/20060206155001/http://www.nd.edu/~networks/gallery.htm)
- Concise, annotated list of graph theory resources for researchers (http://www.babelgraph.org/links.html)

# Graph (mathematics)

Further information: Graph theory

In mathematics, a **graph** is an abstract representation of a set of objects where some pairs of the objects are connected by links. The interconnected objects are represented by mathematical abstractions called *vertices*, and the links that connect some pairs of vertices are called *edges*. Typically, a graph is depicted in diagrammatic form as a set of dots for the vertices, joined by lines or curves for the edges. Graphs are one of the objects of study in discrete mathematics.

The edges may be directed (asymmetric) or undirected (symmetric). For example, if the vertices represent people at a party, and there is an edge between two

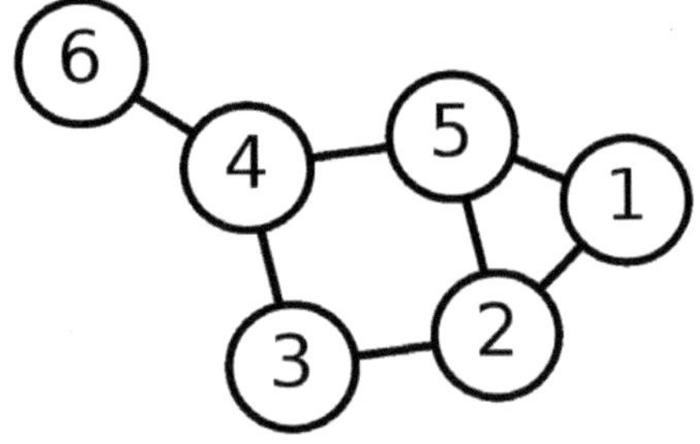

A drawing of a labeled graph on 6 vertices and 7 edges.

people if they shake hands, then this is an undirected graph, because if person A shook hands with person B, then person B also shook hands with person A. On the other hand, if the vertices represent people at a party, and there is an edge from person A to person B when person A knows of person B, then this graph is directed, because knowing of someone is not necessarily a symmetric relation (that is, one person knowing of another person does not necessarily imply the reverse; for example, many fans may know of a celebrity, but the celebrity is unlikely to know of all their fans). This latter type of graph is called a *directed* graph and the edges are called *directed edges* or *arcs*.

Vertices are also called *nodes* or *points*, and edges are also called *lines*. Graphs are the basic subject studied by graph theory. The word "graph" was first used in this sense by J.J. Sylvester in 1878.[1]

# Definitions

Definitions in graph theory vary. The following are some of the more basic ways of defining graphs and related mathematical structures.

## Graph

In the most common sense of the term,[2] a **graph** is an ordered pair $G = (V, E)$ comprising a set $V$ of **vertices** or **nodes** together with a set $E$ of **edges** or **lines**, which are 2-element subsets of $V$ (i.e., an edge is related with two vertices, and the relation is represented as unordered pair of the vertices with respect to the particular edge). To avoid ambiguity, this type of graph may be described precisely as undirected and simple.

Other senses of *graph* stem from different conceptions of the edge set. In one more generalized notion,[3] $E$ is a set together with a relation of **incidence** that associates with each edge two vertices. In another generalized notion, $E$ is a multiset of unordered pairs of (not necessarily distinct) vertices. Many authors call this type of object a multigraph or pseudograph.

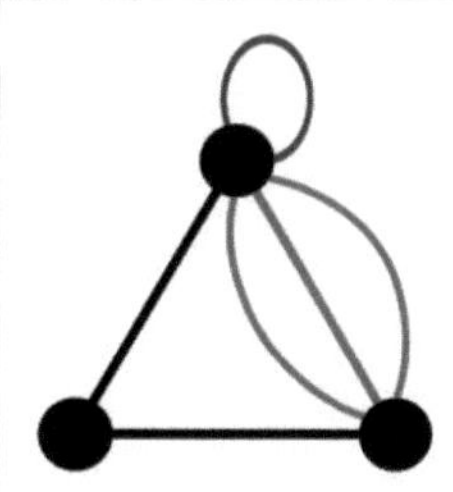

A general example of a graph (actually, a pseudograph) with three vertices and six edges.

All of these variants and others are described more fully below.

The vertices belonging to an edge are called the **ends**, **endpoints**, or **end vertices** of the edge. A vertex may exist in a graph and not belong to an edge.

$V$ and $E$ are usually taken to be finite, and many of the well-known results are not true (or are rather different) for **infinite graphs** because many of the arguments fail in the infinite case. The **order** of a graph is (the number of vertices). A graph's **size** is , the number of edges. The **degree** of a vertex is the number of edges that connect to it, where an edge that connects to the vertex at both ends (a loop) is counted twice.

For an edge $\{u, v\}$, graph theorists usually use the somewhat shorter notation $uv$.

## Adjacency relation

The edges $E$ of an undirected graph $G$ induce a symmetric binary relation ~ on $V$ that is called the **adjacency** relation of $G$. Specifically, for each edge $\{u, v\}$ the vertices $u$ and $v$ are said to be **adjacent** to one another, which is denoted $u \sim v$.

# Types of graphs

## Distinction in terms of the main definition

As stated above, in different contexts it may be useful to define the term *graph* with different degrees of generality. Whenever it is necessary to draw a strict distinction, the following terms are used. Most commonly, in modern texts in graph theory, unless stated otherwise, *graph* means "undirected simple finite graph" (see the definitions below).

### Undirected graph

An undirected graph is one in which edges have no orientation. The edge (a, b) is identical to the edge (b, a), i.e., they are not ordered pairs, but sets $\{u, v\}$ (or 2-multisets) of vertices.

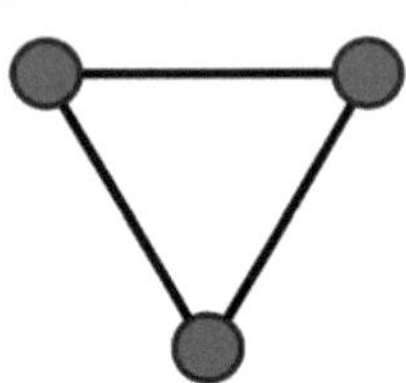

A simple undirected graph with three vertices and three edges. Each vertex has degree two, so this is also a regular graph.

### Directed graph

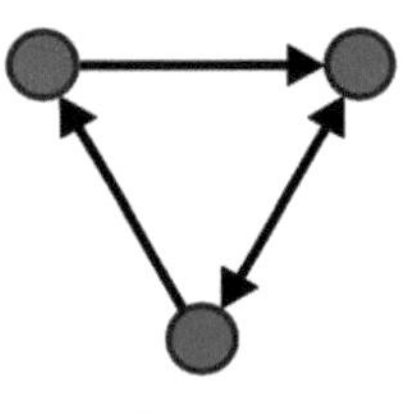

A directed graph

A **directed graph** or **digraph** is an ordered pair $D = (V, A)$ with

- $V$ a set whose elements are called **vertices** or **nodes**, and
- $A$ a set of ordered pairs of vertices, called **arcs**, **directed edges**, or **arrows**.

An arc $a = (x, y)$ is considered to be directed **from** $x$ **to** $y$; $y$ is called the **head** and $x$ is called the **tail** of the arc; $y$ is said to be a **direct successor** of $x$, and $x$ is said to be a **direct predecessor** of $y$. If a path leads from $x$ to $y$, then $y$ is said to be a **successor** of $x$ and **reachable** from $x$, and $x$ is said to be a **predecessor** of $y$. The arc $(y, x)$ is called the arc $(x, y)$ **inverted**.

A directed graph $D$ is called **symmetric** if, for every arc in $D$, the corresponding inverted arc also belongs to $D$. A symmetric loopless directed graph $D = (V, A)$ is equivalent to a simple undirected graph $G = (V, E)$, where the pairs of inverse arcs in $A$ correspond 1-to-1 with the edges in $E$; thus the edges in $G$ number $|E| = |A|/2$, or half the number of arcs in $D$.

A variation on this definition is the **oriented graph**, in which not more than one of $(x, y)$ and $(y, x)$ may be arcs.

### Mixed graph

A **mixed graph** $G$ is a graph in which some edges may be directed and some may be undirected. It is written as an ordered triple $G = (V, E, A)$ with $V$, $E$, and $A$ defined as above. Directed and undirected graphs are special cases.

### Multigraph

A loop is an edge (directed or undirected) which starts and ends on the same vertex; these may be permitted or not permitted according to the application. In this context, an edge with two different ends is called a **link**.

The term "multigraph" is generally understood to mean that multiple edges (and sometimes loops) are allowed. Where graphs are defined so as to *allow* loops and multiple edges, a multigraph is often defined to mean a graph *without* loops,[4] however, where graphs are defined so as to *disallow* loops and multiple edges, the term is often defined to mean a "graph" which can have both multiple edges *and* loops,[5] although many use the term "pseudograph" for this meaning.[6]

### Simple graph

As opposed to a multigraph, a simple graph is an undirected graph that has no loops and no more than one edge between any two different vertices. In a simple graph the edges of the graph form a set (rather than a multiset) and each edge is a *distinct* pair of vertices. In a simple graph with $n$ vertices every vertex has a degree that is less than $n$ (the converse, however, is not true - there exist non-simple graphs with $n$ vertices in which every vertex has a degree smaller than $n$).

### Weighted graph

A graph is a weighted graph if a number (weight) is assigned to each edge. Such weights might represent, for example, costs, lengths or capacities, etc. depending on the problem at hand.

### Half-edges, loose edges

In exceptional situations it is even necessary to have edges with only one end, called **half-edges**, or no ends (**loose edges**); see for example signed graphs and biased graphs.

## Important graph classes

### Regular graph

A regular graph is a graph where each vertex has the same number of neighbors, i.e., every vertex has the same degree or valency. A regular graph with vertices of degree $k$ is called a $k$-regular graph or regular graph of degree $k$.

### Complete graph

Complete graphs have the feature that each pair of vertices has an edge connecting them.

### Finite and infinite graphs

A finite graph is a graph $G = (V, E)$ such that $V$ and $E$ are finite sets. An infinite graph is one with an infinite set of vertices or edges or both.

Most commonly in graph theory it is implied that the graphs discussed are finite. If the graphs are infinite, that is usually specifically stated.

A complete graph with 5 vertices.
Each vertex has an edge to every other vertex.

### Graph classes in terms of connectivity

In an undirected graph $G$, two vertices $u$ and $v$ are called **connected** if $G$ contains a path from $u$ to $v$. Otherwise, they are called **disconnected**. A graph is called **connected** if every pair of distinct vertices in the graph is connected; otherwise, it is called **disconnected**.

A graph is called ***k*-vertex-connected** or ***k*-edge-connected** if no set of *k-1* vertices (respectively, edges) exists that, when removed, disconnects the graph. A $k$-vertex-connected graph is often called simply ***k*-connected**.

A directed graph is called **weakly connected** if replacing all of its directed edges with undirected edges produces a connected (undirected) graph. It is **strongly connected** or **strong** if it contains a directed path from $u$ to $v$ and a directed path from $v$ to $u$ for every pair of vertices $u$, $v$.

## Properties of graphs

Two edges of a graph are called **adjacent** (sometimes **coincident**) if they share a common vertex. Two arrows of a directed graph are called **consecutive** if the head of the first one is at the nock (notch end) of the second one. Similarly, two vertices are called **adjacent** if they share a common edge (**consecutive** if they are at the notch and at the head of an arrow), in which case the common edge is said to **join** the two vertices. An edge and a vertex on that edge are called **incident**.

The graph with only one vertex and no edges is called the **trivial graph**. A graph with only vertices and no edges is known as an **edgeless graph**. The graph with no vertices and no edges is sometimes called the **null graph** or **empty graph**, but the terminology is not consistent and not all mathematicians allow this object.

In a **weighted** graph or digraph, each edge is associated with some value, variously called its *cost*, *weight*, *length* or other term depending on the application; such graphs arise in many contexts, for example in optimal routing problems such as the traveling salesman problem.

Normally, the vertices of a graph, by their nature as elements of a set, are distinguishable. This kind of graph may be called **vertex-labeled**. However, for many questions it is better to treat vertices as indistinguishable; then the graph may be called **unlabeled**. (Of course, the vertices may be still distinguishable by the properties of the graph itself, e.g., by the numbers of incident edges). The same remarks apply to edges, so graphs with labeled edges are called **edge-labeled** graphs. Graphs with labels attached to edges or vertices are more generally designated as **labeled**. Consequently, graphs in which vertices are indistinguishable and edges are indistinguishable are called *unlabeled*. (Note that in the literature the term *labeled* may apply to other kinds of labeling, besides that which serves only to distinguish different vertices or edges.)

## Examples

- The diagram at right is a graphic representation of the following graph:

    $V = \{1, 2, 3, 4, 5, 6\}$

    $E = .$

- In category theory a small category can be represented by a directed multigraph in which the objects of the category represented as vertices and the morphisms as directed edges. Then, the functors between categories induce some, but not necessarily all, of the digraph morphisms of the graph.

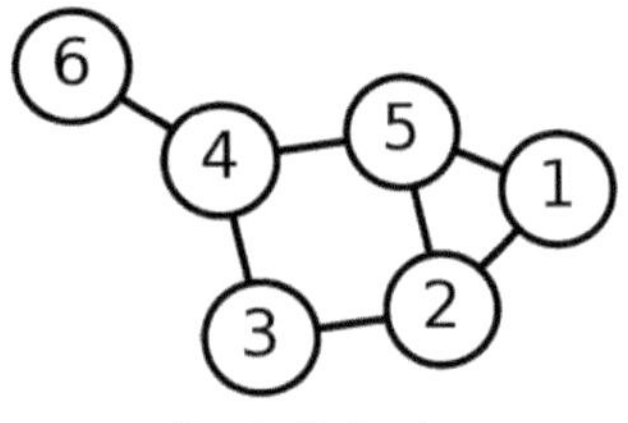

A graph with six nodes.

- In computer science, directed graphs are used to represent knowledge (e.g., Conceptual graph), finite state machines, and many other discrete structures.

- A binary relation $R$ on a set $X$ defines a directed graph. An element $x$ of $X$ is a direct predecessor of an element $y$ of $X$ iff $xRy$.

## Important graphs

Basic examples are:

- In a complete graph, each pair of vertices is joined by an edge; that is, the graph contains all possible edges.
- In a bipartite graph, the vertex set can be partitioned into two sets, $W$ and $X$, so that no two vertices in $W$ are adjacent and no two vertices in $X$ are adjacent. Alternatively, it is a graph with a chromatic number of 2.
- In a complete bipartite graph, the vertex set is the union of two disjoint sets, $W$ and $X$, so that every vertex in $W$ is adjacent to every vertex in $X$ but there are no edges within $W$ or $X$.
- In a *linear graph* or path graph of length $n$, the vertices can be listed in order, $v_0$, $v_1$, ..., $v_n$, so that the edges are $v_{i-1}v_i$ for each $i = 1, 2, ..., n$. If a linear graph occurs as a subgraph of another graph, it is a path in that graph.
- In a cycle graph of length $n \geq 3$, vertices can be named $v_1$, ..., $v_n$ so that the edges are $v_{i-1}v_i$ for each $i = 2,...,n$ in addition to $v_n v_1$. Cycle graphs can be characterized as connected 2-regular graphs. If a cycle graph occurs as a subgraph of another graph, it is a *cycle* or *circuit* in that graph.
- A planar graph is a graph whose vertices and edges can be drawn in a plane such that no two of the edges intersect (i.e., *embedded* in a plane).
- A tree is a connected graph with no cycles.
- A *forest* is a graph with no cycles (i.e. the disjoint union of one or more *trees*).

More advanced kinds of graphs are:

- The Petersen graph and its generalizations
- Perfect graphs
- Cographs
- Chordal graphs
- Other graphs with large automorphism groups: vertex-transitive, arc-transitive, and distance-transitive graphs.
- Strongly regular graphs and their generalization distance-regular graphs.

## Operations on graphs

There are several operations that produce new graphs from old ones, which might be classified into the following categories:

- Elementary operations, sometimes called "editing operations" on graphs, which create a new graph from the original one by a simple, local change, such as addition or deletion of a vertex or an edge, merging and splitting of vertices, etc.
- Graph rewrite operations replacing the occurrence of some pattern graph within the host graph by an instance of the corresponding replacement graph.
- Unary operations, which create a significantly new graph from the old one. Examples:
  - Line graph
  - Dual graph
  - Complement graph
- Binary operations, which create new graph from two initial graphs. Examples:
  - Disjoint union of graphs
  - Cartesian product of graphs
  - Tensor product of graphs
  - Strong product of graphs
  - Lexicographic product of graphs

## Generalizations

In a hypergraph, an edge can join more than two vertices.

An undirected graph can be seen as a simplicial complex consisting of 1-simplices (the edges) and 0-simplices (the vertices). As such, complexes are generalizations of graphs since they allow for higher-dimensional simplices.

Every graph gives rise to a matroid.

In model theory, a graph is just a structure. But in that case, there is no limitation on the number of edges: it can be any cardinal number, see continuous graph.

In computational biology, power graph analysis introduces power graphs as an alternative representation of undirected graphs.

In geographic information systems, geometric networks are closely modeled after graphs, and borrow many concepts from graph theory to perform spatial analysis on road networks or utility grids.

## See also

- Dual graph
- Glossary of graph theory
- Graph (data structure)
- Graph database
- Graph drawing
- Graph theory publications
- List of graph theory topics
- Network theory
- Webgraph
- Conceptual graph
- Horizontal constraint graph
- Causal dynamical triangulation
- Sage Math (software)
- NetworkX (software)
- Mathematica (software)

## Notes

[1]  Gross, Jonathan L.; Yellen, Jay (2004). *Handbook of graph theory* (http://books.google.com/?id=mKkIGIea_BkC). CRC Press. p. 35
       (http://books.google.com/books?id=mKkIGIea_BkC&pg=PA35&lpg=PA35). ISBN 9781584880905.
[2]  See, for instance, Iyanaga and Kawada, **69 J**, p. 234 or Biggs, p. 4.
[3]  See, for instance, Graham et al., p. 5.
[4]  For example, see Balakrishnan, p. 1, Gross (2003), p. 4, and Zwillinger, p. 220.
[5]  For example, see. Bollobás, p. 7 and Diestel, p. 25.
[6]  Gross (1998), p. 3, Gross (2003), p. 205, Harary, p.10, and Zwillinger, p. 220.

# References


- Balakrishnan, V. K. (1997-02-01). *Graph Theory* (1st ed.). McGraw-Hill. ISBN 0-07-005489-4.
- Berge, Claude (1958) (in French). *Théorie des graphes et ses applications*. Dunod, Paris: Collection Universitaire de Mathématiques, II. pp. viii+277. Translation: . Dover, New York: Wiley. 2001 [1962].
- Biggs, Norman (1993). *Algebraic Graph Theory* (2nd ed.). Cambridge University Press. ISBN 0-521-45897-8.
- Bollobás, Béla (2002-08-12). *Modern Graph Theory* (1st ed.). Springer. ISBN 0-387-98488-7.
- Bang-Jensen, J.; Gutin, G. (2000). *Digraphs: Theory, Algorithms and Applications* (http://www.cs.rhul.ac.uk/books/dbook/). Springer.
- Diestel, Reinhard (2005). *Graph Theory* (http://diestel-graph-theory.com/GrTh.html) (3rd ed.). Berlin, New York: Springer-Verlag. ISBN 978-3-540-26183-4.
- Graham, R.L., Grötschel, M., and Lovász, L, ed (1995). *Handbook of Combinatorics*. MIT Press. ISBN 0-262-07169-X.
- Gross, Jonathan L.; Yellen, Jay (1998-12-30). *Graph Theory and Its Applications*. CRC Press. ISBN 0-8493-3982-0.
- Gross, Jonathan L., & Yellen, Jay, ed (2003-12-29). *Handbook of Graph Theory*. CRC. ISBN 1-58488-090-2.
- Harary, Frank (January 1995). *Graph Theory*. Addison Wesley Publishing Company. ISBN 0-201-41033-8.
- Iyanaga, Shôkichi; Kawada, Yukiyosi (1977). *Encyclopedic Dictionary of Mathematics*. MIT Press. ISBN 0-262-09016-3.
- Zwillinger, Daniel (2002-11-27). *CRC Standard Mathematical Tables and Formulae* (31st ed.). Chapman & Hall/CRC. ISBN 1-58488-291-3.


# External links

- A searchable database of small connected graphs (http://www.gfredericks.com/main/sandbox/graphs)
- VisualComplexity.com (http://www.visualcomplexity.com) — A visual exploration on mapping complex networks
- Weisstein, Eric W., " Graph (http://mathworld.wolfram.com/Graph.html)" from MathWorld.
- Intelligent Graph Visualizer (https://sourceforge.net/projects/igv-intelligent/) — IGV create and edit graph, automatically places graph, search shortest path (+coloring vertices), center, degree, eccentricity, etc.
- Visual Graph Editor 2 (http://code.google.com/p/vge2/) — VGE2 designed for quick and easy creation, editing and saving of graphs and analysis of problems connected with graphs.
- GraphsJ 2 (http://gianlucacosta.altervista.org/software/graphsj2/index.php) — GraphsJ is a didactic Java software which features an easy-to-use GUI and interactively solves step-by-step many graph problems executing their associated algorithms.

# Edge-transitive graph

| Graph families defined by their automorphisms | | |
|---|---|---|
| distance-transitive | distance-regular | strongly regular |
| symmetric (arc-transitive) | $t$-transitive, $t \geq 2$ | |
| (if connected) | | |
| vertex- and edge-transitive | edge-transitive and regular | edge-transitive |
| vertex-transitive | regular | |
| Cayley graph | skew-symmetric | asymmetric |

In the mathematical field of graph theory, an **edge-transitive graph** is a graph $G$ such that, given any two edges $e_1$ and $e_2$ of $G$, there is an automorphism of $G$ that maps $e_1$ to $e_2$.[1]

In other words, a graph is edge-transitive if its automorphism group acts transitively upon its edges.

## Examples and properties

Edge-transitive graphs include any complete bipartite graph , and any symmetric graph, such as the vertices and edges of the cube.[1] Symmetric graphs are also vertex-transitive (if they are connected), but in general edge-transitive graphs need not be vertex-transitive. The Gray graph is an example of a graph which is edge-transitive but not vertex-transitive. All such graphs are bipartite,[1] and hence can be colored with only two colors.

An edge-transitive graph that is also regular, but not vertex-transitive, is called semi-symmetric. The Gray graph again provides an example.

The Gray graph is edge-transitive and regular, but not vertex-transitive.

## See also

- Edge-transitive (in geometry)

## References

[1]  Biggs, Norman (1993). *Algebraic Graph Theory* (2nd ed. ed.). Cambridge: Cambridge University Press. p. 118. ISBN 0-521-45897-8.

## External links

- Weisstein, Eric W., " Edge-transitive graph (http://mathworld.wolfram.com/Edge-TransitiveGraph.html)" from MathWorld.

# Vertex-transitive graph

| Graph families defined by their automorphisms | | |
|---|---|---|
| distance-transitive | distance-regular | strongly regular |
| symmetric (arc-transitive) | $t$-transitive, $t \geq 2$ | |
| (if connected) | | |
| vertex- and edge-transitive | edge-transitive and regular | edge-transitive |
| vertex-transitive | regular | |
| Cayley graph | skew-symmetric | asymmetric |

In the mathematical field of graph theory, a **vertex-transitive graph** is a graph $G$ such that, given any two vertices $v_1$ and $v_2$ of $G$, there is some automorphism

such that

In other words, a graph is vertex-transitive if its automorphism group acts transitively upon its vertices.[1] A graph is vertex-transitive if and only if its graph complement is, since the group actions are identical.

Every symmetric graph without isolated vertices is vertex-transitive, and every vertex-transitive graph is regular. However, not all vertex-transitive graphs are symmetric (for example, the edges of the truncated tetrahedron), and not all regular graphs are vertex-transitive (for example, the Frucht graph).

## Finite examples

Finite vertex-transitive graphs include the symmetric graphs (such as the Petersen graph, the Heawood graph and the vertices and edges of the Platonic solids). The finite Cayley graphs (such as cube-connected cycles) are also vertex-transitive, as are the vertices and edges of the Archimedean solids (though only two of these are symmetric).

## Properties

The edge-connectivity of a vertex-transitive graph is equal to the degree $d$, while the vertex-connectivity will be at least $2(d+1)/3$.[2] If the degree is 4 or less, or the graph is also edge-transitive, or the graph is a minimal Cayley graph, then the vertex-connectivity will also be equal to $d$.[3]

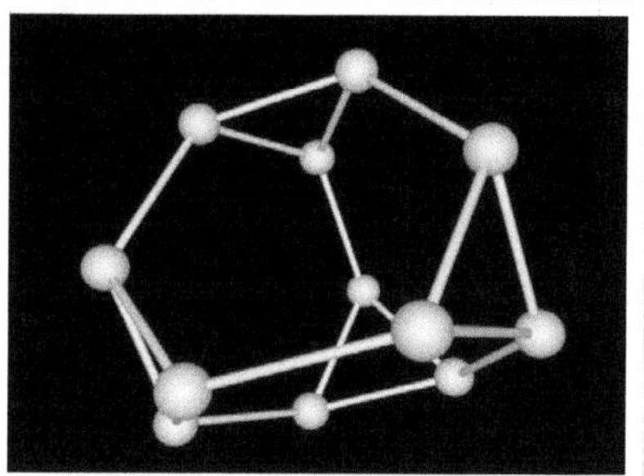

The edges of the truncated tetrahedron form a vertex-transitive graph (also a Cayley graph) which is not symmetric.

## Infinite examples

Infinite vertex-transitive graphs include:

- infinite paths (infinite in both directions)
- infinite regular trees, e.g. the Cayley graph of the free group
- graphs of uniform tessellations (see a complete list of planar tessellations), including all tilings by regular polygons
- infinite Cayley graphs

- the Rado graph

Two countable vertex-transitive graphs are called quasi-isometric if the ratio of their distance functions is bounded from below and from above. A well known conjecture states that every infinite vertex-transitive graph is quasi-isometric to a Cayley graph. A counterexample has been proposed by Diestel and Leader.[4] Most recently, Eskin, Fisher, and Whyte confirmed the counterexample.[5]

## See also

- Edge-transitive graph
- Lovász conjecture
- Semi-symmetric graph

## References

[1] Godsil, Chris; Royle, Gordon (2001), *Algebraic Graph Theory*, Graduate Texts in Mathematics, **207**, New York: Springer-Verlag.

[2] Godsil, C. and Royle, G. (2001), *Algebraic Graph Theory*, Springer Verlag

[3] Babai, L. (1996), *Technical Report TR-94-10*, University of Chicago (http://www.cs.uchicago.edu/files/tr_authentic/TR-94-10.ps)

[4] Diestel, Reinhard; Leader, Imre (2001), "A conjecture concerning a limit of non-Cayley graphs" (http://www.math.uni-hamburg.de/home/diestel/papers/Cayley.pdf), *Journal of Algebraic Combinatorics* **14** (1): 17–25, doi:10.1023/A:1011257718029, .

[5] Eskin, Alex; Fisher, David; Whyte, Kevin (2005). "Quasi-isometries and rigidity of solvable groups". arXiv:math.GR/0511647..

# Graph automorphism

In the mathematical field of graph theory, an **automorphism** of a graph is a form of symmetry in which the graph is mapped onto itself while preserving the edge–vertex connectivity.

Formally, an automorphism of a graph $G = (V,E)$ is a permutation $\sigma$ of the vertex set $V$, such that the pair of vertices $(u,v)$ form an edge if and only if the pair $(\sigma(u),\sigma(v))$ also form an edge. That is, it is a graph isomorphism from $G$ to itself. Automorphisms may be defined in this way both for directed graphs and for undirected graphs. The composition of two automorphisms is another automorphism, and the set of automorphisms of a given graph, under the composition operation, forms a group, the automorphism group of the graph. In the opposite direction, by Frucht's theorem, all groups can be represented as the automorphism group of a connected graph − indeed, of a cubic graph.[1] [2]

# Computational complexity

Constructing the automorphism group is at least as difficult (in terms of its computational complexity) as solving the graph isomorphism problem, determining whether two given graphs correspond vertex-for-vertex and edge-for-edge. For, $G$ and $H$ are isomorphic if and only if the disconnected graph formed by the disjoint union of graphs $G$ and $H$ has an automorphism that swaps the two components.[3]

The **graph automorphism problem** is the problem of testing whether a graph has a nontrivial automorphism. It belongs to the class NP of computational complexity. Similar to the graph isomorphism problem, it is unknown whether it has a polynomial time algorithm or it is NP-complete.[4] There is a polynomial time algorithm for solving the graph automorphism problem for graphs where vertex degrees are bounded by a constant (Luks 1982). It is known that the graph automorphism problem is polynomial-time many-one reducible to the graph isomorphism problem, but the converse reduction is unknown.[5] [6] [7]

This drawing of the Petersen graph displays a subgroup of its symmetries, isomorphic to the dihedral group $D_5$, but the graph has additional symmetries that are not present in the drawing (since the graph is symmetric, all links are equivalent, for example).

# Algorithms, software and applications

While no *worst-case* polynomial-time algorithms are known for the general Graph Automorphism problem, finding the automorphism group (and printing out an irredundant set of generators) for many large graphs arising in applications is rather easy. Several open-source software tools are available for this task, including NAUTY [8][9], BLISS [10][11] and SAUCY [12][13] [14]. SAUCY and BLISS are particularly efficient for sparse graphs, e.g., SAUCY processes some graphs with millions of vertices in mere seconds. However, BLISS and NAUTY can also produce Canonical Labeling, whereas SAUCY is currently optimized for solving Graph Automorphism.

Practical applications of Graph Automorphism include graph drawing and other visualization tasks, solving structured instances of Boolean Satisfiability arising in the context of Formal verification and Logistics. Molecular symmetry can predict or explain chemical properties.

# Symmetry display

Several graph drawing researchers have investigated algorithms for drawing graphs in such a way that the automorphisms of the graph become visible as symmetries of the drawing. This may be done either by using a method that is not designed around symmetries, but that automatically generates symmetric drawings when possible,[15] or by explicitly identifying symmetries and using them to guide vertex placement in the drawing.[16] It is not always possible to display all symmetries of the graph simultaneously, so it may be necessary to choose which symmetries to display and which to leave unvisualized.

# Graph families defined by their automorphisms

Several families of graphs are defined by having certain types of automorphisms:

- An asymmetric graph is an undirected graph without any nontrivial automorphisms.
- A vertex-transitive graph is an undirected graph in which every vertex may be mapped by an automorphism into any other vertex.
- An edge-transitive graph is an undirected graph in which every edge may be mapped by an automorphism into any other edge.
- A symmetric graph is a graph such that every pair of adjacent vertices may be mapped by an automorphism into any other pair of adjacent vertices.
- A distance-transitive graph is a graph such that every pair of vertices may be mapped by an automorphism into any other pair of vertices that are the same distance apart.
- A semi-symmetric graph is a graph that is edge-transitive but not vertex-transitive.
- A half-transitive graph is a graph that is vertex-transitive and edge-transitive but not symmetric.
- A skew-symmetric graph is a directed graph together with a permutation σ on the vertices that maps edges to edges but reverses the direction of each edge. Additionally, σ is required to be an involution.

Inclusion relationships between these families are indicated by the following table:

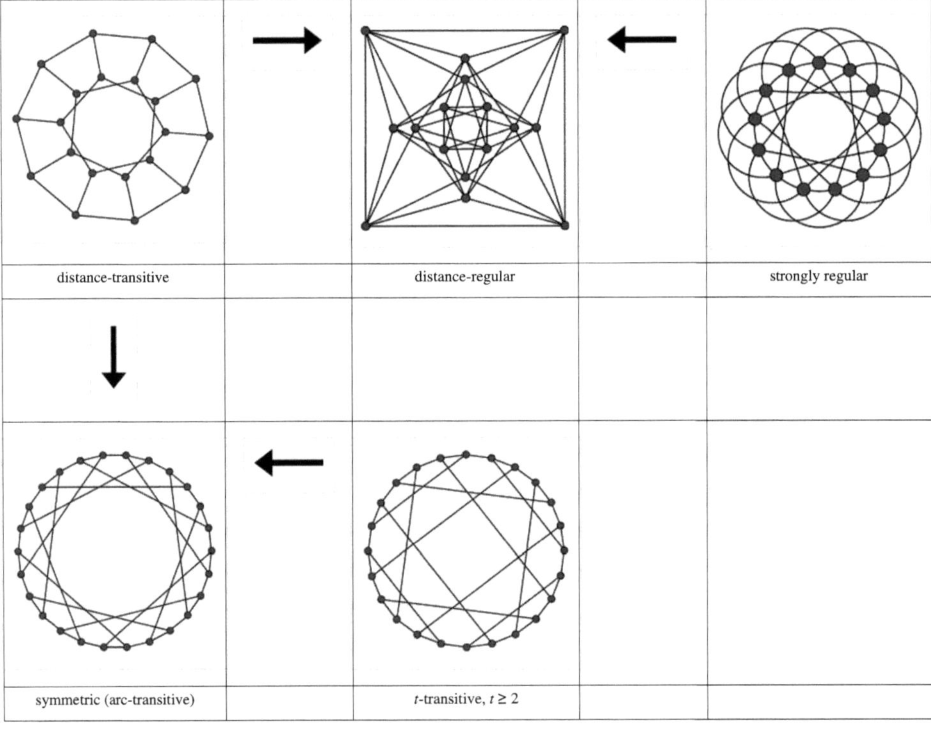

| distance-transitive | | distance-regular | | strongly regular |
| --- | --- | --- | --- | --- |
| symmetric (arc-transitive) | | $t$-transitive, $t \geq 2$ | | |

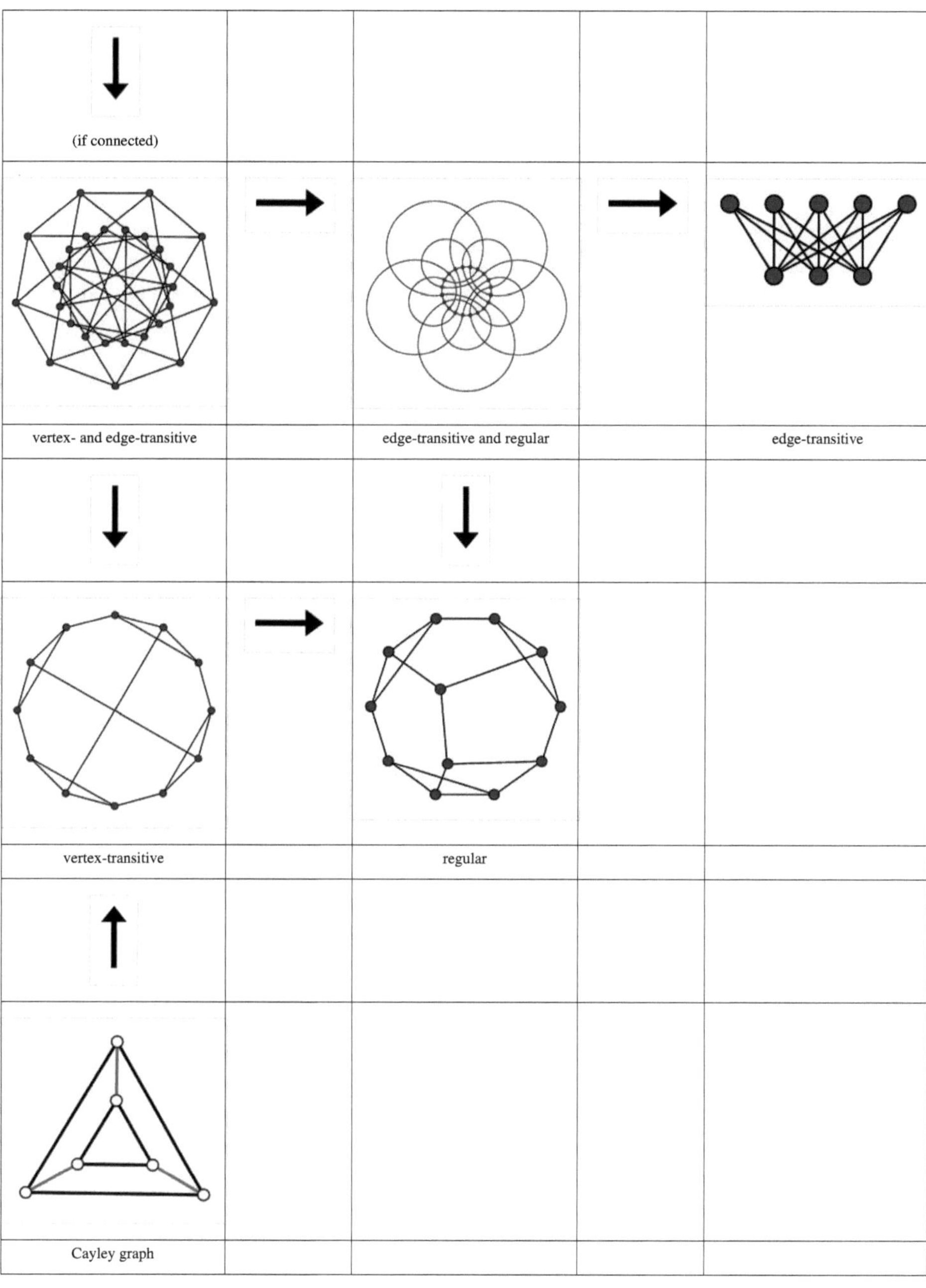

(if connected)
vertex- and edge-transitive
edge-transitive and regular
edge-transitive
vertex-transitive
regular
Cayley graph

## See also

* Algebraic graph theory

## References

[1] Frucht, R. (1938), "Herstellung von Graphen mit vorgegebener abstrakter Gruppe" (http://www.numdam.org/
item?id=CM_1939__6__239_0) (in German), *Compositio Mathematica* **6**: 239–250, ISSN 0010-437X, Zbl 0020.07804, .

[2] Frucht, R. (1949), "Graphs of degree three with a given abstract group" (http://cms.math.ca/cjm/v1/p365), *Canadian Journal of
Mathematics* **1** (4): 365–378, doi:10.4153/CJM-1949-033-6, ISSN 0008-414X, MR0032987, .

[3] Luks, Eugene M. (1982), "Isomorphism of graphs of bounded valence can be tested in polynomial time", *Journal of Computer and System
Sciences* **25** (1): 42–65, doi:10.1016/0022-0000(82)90009-5.

[4] A. Lubiw, "Some NP-complete problems similar to Graph Isomorphism", SIAM Journal on Computing, 1O.ll-21, 1981.

[5] R. Mathon, "A note on the graph isomorphism counting problem", *Information Processing Letters,* 8 (1979) pp. 131-132

[6] Köbler, Johannes; Uwe Schöning, Jacobo Torán (1993), *Graph Isomorphism Problem: The Structural Complexity,* Birkhäuser Verlag,
ISBN 0817636803, OCLC 246882287

[7] Jacobo Torán, " On the Hardness of Graph Isomorphism (http://theorie.informatik.uni-ulm.de/Personen/toran/papers/hard.pdf)", *SIAM
Journal on Computing,* vol. 33, no. 5, 2004, pp. 1093-1108, doi:10.1137/S009753970241096X

[8] http://cs.anu.edu.au/people/bdm/nauty/

[9] McKay, Brendan (1981), "Practical Graph Isomorphism" (http://cs.anu.edu.au/people/bdm/nauty/pgi.pdf), *Congressus Numerantium*
**30**: 45–87, , retrieved 14 April 2011.

[10] http://www.tcs.hut.fi/Software/bliss/

[11] Junttila, Tommi; Kaski, Petteri (2007), "Engineering an efficient canonical labeling tool for large and sparse graphs" (http://www.siam.
org/proceedings/alenex/2007/alx07_013junttilat.pdf), *Proceedings of the Ninth Workshop on Algorithm Engineering and Experiments
(ALENEX07),* .

[12] http://vlsicad.eecs.umich.edu/BK/SAUCY/

[13] Darga, Paul; Sakallah, Karem; Markov, Igor L. (June 2008), "Faster Symmetry Discovery using Sparsity of Symmetries" (http://vlsicad.
eecs.umich.edu/BK/SAUCY/saucy-dac08.pdf), *Proceedings of the 45st Design Automation Conference*: 149–154,
doi:10.1145/1391469.1391509, ISBN 9781605581156, .

[14] Katebi, Hadi; Sakallah, Karem; Markov, Igor L. (July 2010), "Symmetry and Satisfiability: An Update" (http://www.eecs.umich.edu/
~imarkov/pubs/conf/sat10-sym.pdf), *Proc. Satisfiability Symposium (SAT),* .

[15] Di Battista, Giuseppe; Tamassia, Roberto; Tollis, Ioannis G. (1992), "Area requirement and symmetry display of planar upward drawings",
*Discrete and Computational Geometry* **7** (1): 381–401, doi:10.1007/BF02187850; Eades, Peter; Lin, Xuemin (2000), "Spring algorithms and
symmetry", *Theoretical Computer Science* **240** (2): 379–405, doi:10.1016/S0304-3975(99)00239-X.

[16] Hong, Seok-Hee (2002), "Drawing Graphs Symmetrically in Three Dimensions", *Proc. 9th Int. Symp. Graph Drawing (GD 2001),* Lecture
Notes in Computer Science, **2265**, Springer-Verlag, pp. 106–108, doi:10.1007/3-540-45848-4_16, ISBN 978-3-540-43309-5.

# Group action

In algebra and geometry, a **group action** is a way of describing symmetries of objects using groups. The essential elements of the object are described by a set, and the symmetries of the object are described by the symmetry group of this set, which consists of bijective transformations of the set. In this case, the group is also called a **permutation group** (especially if the set is finite or not a vector space) or **transformation group** (especially if the set is a vector space and the group acts like linear transformations of the set).

A group action is an extension to the definition of a symmetry group in which every element of the group "acts" like a bijective transformation (or "symmetry") of some set, without being identified with that transformation. This allows for a more comprehensive description of the symmetries of an object, such as a polyhedron, by allowing the same group to act on several different sets of features, such as the set of vertices, the set of edges and the set of faces of the polyhedron.

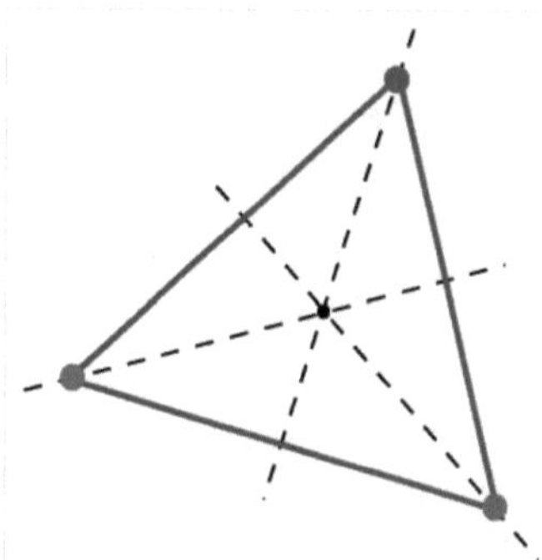

Given an equilateral triangle, the counterclockwise rotation by 120° around the center of the triangle "acts" on the set of vertices of the triangle by mapping every vertex to another one.

If $G$ is a group and $X$ is a set then a group action may be defined as a group homomorphism from $G$ to the symmetric group of $X$. The action assigns a permutation of $X$ to each element of the group in such a way that the permutation of $X$ assigned to:

- The identity element of $G$ is the identity transformation of $X$;
- A product $gh$ of two elements of $G$ is the composite of the permutations assigned to $g$ and $h$.

Since each element of $G$ is represented as a permutation, a group action is also known as a **permutation representation**.

The abstraction provided by group actions is a powerful one, because it allows geometrical ideas to be applied to more abstract objects. Many objects in mathematics have natural group actions defined on them. In particular, groups can act on other groups, or even on themselves. Despite this generality, the theory of group actions contains wide-reaching theorems, such as the orbit stabilizer theorem, which can be used to prove deep results in several fields.

## Definition

If is a group and is a set, then a (*left*) *group action* of $G$ on $X$ is a binary operator:

that satisfies the following two axioms:[1]

Associativity

;

Identity

.

The set $X$ is called a (*left*) *G-set*. The group $G$ is said to act on $X$ (on the left).

From these two axioms, it follows that for every $g$ in $G$, the function which maps $x$ in X to $g{\cdot}x$ is a bijective map from $X$ to $X$ (its inverse being the function which maps $x$ to $g^{-1}{\cdot}x$). Therefore, one may alternatively define a group action of $G$ on $X$ as a group homomorphism from $G$ into the symmetric group $Sym(X)$ of all bijections from $X$ to $X$.[2]

In complete analogy, one can define a *right group action* of $G$ on $X$ as a function $X \times G \to X$ by the two axioms:

Associativity

;

Identity

.

The difference between left and right actions is in the order in which a product like $gh$ acts on $x$. For a left action $h$ acts first and is followed by $g$, while for a right action $g$ acts first and is followed by $h$. From a right action a left action can be constructed by composing with the inverse operation on the group. If is a right action, then, a the following is a left action:

Satisfying associativity,

and identity:

Any right action has an equivalent left action, thus only left actions can be considered without any loss of generality. Also, a right action of a group on is the same thing as a left action of its opposite group on .

# Examples

- The *trivial* action for any group $G$ is defined by $g{\cdot}x=x$ for all $g$ in $G$ and all $x$ in $X$; that is, the whole group $G$ induces the identity permutation on $X$.[3]
- Every group $G$ acts on $G$ in two natural but essentially different ways: $g{\cdot}x = gx$ for all $x$ in $G$, or $g{\cdot}x = gxg^{-1}$ for all $x$ in $G$. The latter action is often called the conjugation action, and an exponential notation is commonly used for the right-action variant: $x^g = g^{-1}xg$; it satisfies $(x^g)^h = x^{gh}$.
- The symmetric group $S_n$ and its subgroups act on the set $\{\,1, \ldots , n\,\}$ by permuting its elements
- The symmetry group of a polyhedron acts on the set of vertices of that polyhedron. It also acts on the set of faces or the set of edges of the polyhedron.
- The symmetry group of any geometrical object acts on the set of points of that object
- The automorphism group of a vector space (or graph, or group, or ring...) acts on the vector space (or set of vertices of the graph, or group, or ring...).
- The general linear group $GL(n, \mathbf{R})$, special linear group $SL(n, \mathbf{R})$, orthogonal group $O(n, \mathbf{R})$, and special orthogonal group $SO(n, \mathbf{R})$ are Lie groups which act on $\mathbf{R}^n$.
- The Galois group of a field extension $E/F$ acts on the bigger field $E$. So does every subgroup of the Galois group.
- The additive group of the real numbers $(\mathbf{R}, +)$ acts on the phase space of "well-behaved" systems in classical mechanics (and in more general dynamical systems): if $t$ is in $\mathbf{R}$ and $x$ is in the phase space, then $x$ describes a state of the system, and $t{\cdot}x$ is defined to be the state of the system $t$ seconds later if $t$ is positive or $-t$ seconds ago if $t$ is negative.
- The additive group of the real numbers $(\mathbf{R}, +)$ acts on the set of real functions of a real variable with $(g{\cdot}f)(x)$ equal to e.g. $f(x + g), f(x) + g, , , ,$ or , but not
- The quaternions with modulus 1, as a multiplicative group, act on $\mathbf{R}^3$: for any such quaternion , the mapping $f(\mathbf{x}) = z\,\mathbf{x}\,z^*$ is a counterclockwise rotation through an angle about an axis $\mathbf{v}$; $-z$ is the same rotation; see quaternions and spatial rotation.
- The isometries of the plane act on the set of 2D images and patterns, such as a wallpaper pattern. The definition can be made more precise by specifying what is meant by image or pattern; e.g., a function of position with values in a set of colors.
- More generally, a group of bijections $g: V \to V$ acts on the set of functions $x: V \to W$ by $(gx)(v) = x(g^{-1}(v))$ (or a restricted set of such functions that is closed under the group action). Thus a group of bijections of space induces a group action on "objects" in it.

## Types of actions

The action of $G$ on $X$ is called

- *Transitive* if $X$ is non-empty and if equivalently

  1. For any $x$, $y$ in X there exists a $g$ in $G$ such that $gx = y$,
  2. $Gx = X$ *for all* $x$ in $X$,
  3. $Gx = X$ *for some* $x$ in $X$.

     Here, $Gx = \{g.x \mid g$ in $G\}$ is the orbit of $x$ under $G$.

  - *Sharply transitive* if that $g$ is unique; it is equivalent to regularity defined below.
- *n-transitive* if $X$ has at least $n$ elements and for any pairwise distinct $x_1, ..., x_n$ and pairwise distinct $y_1, ..., y_n$ there is a $g$ in $G$ such that $g.x_k = y_k$ for $1 \leq k \leq n$. A 2-transitive action is also called *doubly transitive*, a 3-transitive action is also called *triply transitive*, and so on. Such actions define 2-transitive groups, 3-transitive groups, and multiply transitive groups.

  - *Sharply n-transitive* if there is exactly one such $g$. See also sharply triply transitive groups.
- *Faithful* (or *effective*) if for any two distinct $g$, $h$ in $G$ there exists an $x$ in $X$ such that $g{\cdot}x \neq h{\cdot}x$; or equivalently, if for any $g \neq e$ in $G$ there exists an $x$ in $X$ such that $g{\cdot}x \neq x$. Intuitively, different elements of G induce different permutations of X.
- *Free* (or *semiregular*) if for any $x$ in $X$, $g.x = h.x$ implies $g = h$. Equivalently: if there exists an $x$ in $X$ such that $g.x = x$ (that is, if $g$ has at least one fixed point), then $g$ is the identity.
- *Regular* (or *simply transitive*) if it is both transitive and free; this is equivalent to saying that for any two $x$, $y$ in $X$ there exists precisely one $g$ in $G$ such that $g{\cdot}x = y$. In this case, $X$ is known as a principal homogeneous space for $G$ or as a G-torsor.
- *Primitive* if it is transitive and preserves no non-trivial partition of $X$. See Primitive permutation group for details.
- *Locally free* if $G$ is a topological group, and there is a neighbourhood $U$ of $e$ in $G$ such that the restriction of the action to $U$ is free; that is, if $g{\cdot}x = x$ for some $x$ and some $g$ in $U$ then $g = e$.
- *Irreducible* if $X$ is a non-zero module over a ring $R$, the action of $G$ is $R$-linear, and there is no nonzero proper invariant submodule.

Every free action on a non-empty set is faithful. A group $G$ acts faithfully on $X$ if and only if the homomorphism $G \rightarrow \mathrm{Sym}(X)$ has a trivial kernel. Thus, for a faithful action, $G$ is isomorphic to a permutation group on $X$; specifically, $G$ is isomorphic to its image in $\mathrm{Sym}(X)$.

The action of any group $G$ on itself by left multiplication is regular, and thus faithful as well. Every group can, therefore, be embedded in the symmetric group on its own elements, $\mathrm{Sym}(G)$ — a result known as Cayley's theorem.

If $G$ does not act faithfully on $X$, one can easily modify the group to obtain a faithful action. If we define $N = \{g$ in $G : g{\cdot}x = x$ for all $x$ in $X\}$, then $N$ is a normal subgroup of $G$; indeed, it is the kernel of the homomorphism $G \rightarrow \mathrm{Sym}(X)$. The factor group $G/N$ acts faithfully on $X$ by setting $(gN){\cdot}x = g{\cdot}x$. The original action of $G$ on $X$ is faithful if and only if $N = \{e\}$.

## Orbits and stabilizers

Consider a group $G$ acting on a set $X$. The *orbit* of a point $x$ in $X$ is the set of elements of $X$ to which $x$ can be moved by the elements of $G$. The orbit of $x$ is denoted by $Gx$:

The defining properties of a group guarantee that the set of orbits of (points $x$ in) $X$ under the action of $G$ form a partition of $X$. The associated equivalence relation is defined by saying $x \sim y$ if and only if there exists a $g$ in $G$ with $g \cdot x = y$. The orbits are then the equivalence classes under this relation; two elements $x$ and $y$ are equivalent if and only if their orbits are the same; i.e., $Gx = Gy$.

In the compound of five tetrahedra, the symmetry group is the (rotational) icosahedral group $I$ of order 60, while the stabilizer of a single chosen tetrahedron is the (rotational) tetrahedral group $T$ of order 12, and the orbit space $I/T$ (of order $60/12 = 5$) is naturally identified with the 5 tetrahedra — the coset $gT$ corresponds to which tetrahedron $g$ sends the chosen tetrahedron to.

The set of all orbits of $X$ under the action of $G$ is written as $X/G$ (or, less frequently: $G \backslash X$), and is called the *quotient* of the action. In geometric situations it may be called the *orbit space*, while in algebraic situations it may be called the space of *coinvariants*, and written by contrast with the invariants (fixed points), denoted the coinvariants are a *quotient* while the invariants are a *subset*. The coinvariant terminology and notation are used particularly in group cohomology and group homology, which use the same superscript/subscript convention.

## Invariant subsets

If $Y$ is a subset of $X$, we write $GY$ for the set $\{\, g \cdot y : y \in Y \text{ and } g \in G \}$. We call the subset $Y$ *invariant under $G$* if $GY = Y$ (which is equivalent to $GY \subseteq Y$). In that case, $G$ also operates on $Y$. The subset $Y$ is called *fixed under $G$* if $g \cdot y = y$ for all $g$ in $G$ and all $y$ in $Y$. Every subset that's fixed under $G$ is also invariant under $G$, but not vice versa.

Every orbit is an invariant subset of $X$ on which $G$ acts transitively. The action of $G$ on $X$ is *transitive* if and only if all elements are equivalent, meaning that there is only one orbit.

## Stabilizer subgroup

For every $x$ in $X$, we define the *stabilizer subgroup* of $x$ (also called the *isotropy group* or *little group*) as the set of all elements in $G$ that fix $x$:

This is a subgroup of $G$, though typically not a normal one. The action of $G$ on $X$ is free if and only if all stabilizers are trivial. The kernel $N$ of the homomorphism $G \to \mathrm{Sym}(X)$ is given by the intersection of the stabilizers $G_x$ for all $x$ in $X$.

A useful result is the following. Let $x$ and $y$ be two distinct elements in $X$, and let $g$ be a group element such that . Then the two isotropy groups and are related by . Let us prove this: by definition if and only if . Applying to both sides of this equality we get ; that is, . This shows that if and only if .

## Orbit-stabilizer theorem

Orbits and stabilizers are closely related. For a fixed $x$ in $X$, consider the map from $G$ to $X$ given by $g \mapsto g \cdot x$. The image of this map is the orbit of $x$ and the coimage is the set of all left cosets of $G_x$. The standard quotient theorem of set theory then gives a natural bijection between $G/G_x$ and $Gx$. Specifically, the bijection is given by $hG_x \mapsto h \cdot x$. This result is known as the *orbit-stabilizer theorem*.

If $G$ and $X$ are finite then the orbit-stabilizer theorem, together with Lagrange's theorem, gives

This result is especially useful since it can be employed for counting arguments.

Note that if two elements $x$ and $y$ belong to the same orbit, then their stabilizer subgroups, $G_x$ and $G_y$, are conjugate (in particular, they are isomorphic). More precisely: if $y = g \cdot x$, then $G_y = gG_x g^{-1}$. Points with conjugate stabilizer subgroups are said to have the same *orbit-type*.

A result closely related to the orbit-stabilizer theorem is Burnside's lemma:

where $X^g$ is the set of points fixed by $g$. This result is mainly of use when $G$ and $X$ are finite, when it can be interpreted as follows: the number of orbits is equal to the average number of points fixed per group element.

The set of formal differences of finite $G$-sets forms a ring called the Burnside ring, where addition corresponds to disjoint union, and multiplication to Cartesian product.

A *G-invariant* element of $X$ is $x \in X$ such that $g \cdot x = x$ for all $g \in G$. The set of all such $x$ is denoted $X^G$ and called the *G-invariants* of $X$. When $X$ is a $G$-module, $X^G$ is the zeroth group cohomology group of $G$ with coefficients in $X$, and the higher cohomology groups are the derived functors of the functor of $G$-invariants.

# Group actions and groupoids

The notion of group action can be put in a broader context by using the associated *action groupoid* associated to the group action, thus allowing techniques from groupoid theory such as presentations and fibrations. Further the stabilisers of the action are the vertex groups, and the orbits of the action are the components, of the action groupoid. For more details, see the book *Topology and groupoids* referenced below.

This action groupoid comes with a morphism which is a *covering morphism of groupoids*. This allows a relation between such morphisms and covering maps in topology.

# Morphisms and isomorphisms between *G*-sets

If $X$ and $Y$ are two $G$-sets, we define a *morphism* from $X$ to $Y$ to be a function $f : X \to Y$ such that $f(g \cdot x) = g \cdot f(x)$ for all $g$ in $G$ and all $x$ in $X$. Morphisms of $G$-sets are also called *equivariant maps* or *G-maps*.

If such a function $f$ is bijective, then its inverse is also a morphism, and we call $f$ an *isomorphism* and the two $G$-sets $X$ and $Y$ are called *isomorphic*; for all practical purposes, they are indistinguishable in this case.

Some example isomorphisms:

- Every regular $G$ action is isomorphic to the action of $G$ on $G$ given by left multiplication.
- Every free $G$ action is isomorphic to $G \times S$, where $S$ is some set and $G$ acts by left multiplication on the first coordinate.
- Every transitive $G$ action is isomorphic to left multiplication by $G$ on the set of left cosets of some subgroup $H$ of $G$.

With this notion of morphism, the collection of all $G$-sets forms a category; this category is a Grothendieck topos (in fact, assuming a classical metalogic, this topos will even be Boolean).

## Continuous group actions

One often considers *continuous group actions*: the group $G$ is a topological group, $X$ is a topological space, and the map $G \times X \to X$ is continuous with respect to the product topology of $G \times X$. The space $X$ is also called a *G-space* in this case. This is indeed a generalization, since every group can be considered a topological group by using the discrete topology. All the concepts introduced above still work in this context, however we define morphisms between $G$-spaces to be *continuous* maps compatible with the action of $G$. The quotient $X/G$ inherits the quotient topology from $X$, and is called the *quotient space* of the action. The above statements about isomorphisms for regular, free and transitive actions are no longer valid for continuous group actions.

If $G$ is a discrete group acting on a topological space $X$, the action is properly discontinuous if for any point $x$ in $X$ there is an open neighborhood $U$ of $x$ in $X$, such that the set of all for which consists of the identity only. If $X$ is a regular covering space of another topological space $Y$, then the action of the deck transformation group on $X$ is properly discontinuous as well as being free. Every free, properly discontinuous action of a group $G$ on a path-connected topological space $X$ arises in this manner: the quotient map $X \mapsto X/G$ is a regular covering map, and the deck transformation group is the given action of $G$ on $X$. Furthermore, if $X$ is simply connected, the fundamental group of will be isomorphic to .

These results have been generalised in the book *Topology and Groupoids* referenced below to obtain the fundamental groupoid of the orbit space of a discontinuous action of a discrete group on a Hausdorff space, as, under reasonable local conditions, the orbit groupoid of the fundamental groupoid of the space. This allows calculations such as the fundamental group of the symmetric square of a space $X$, namely the orbit space of the product of $X$ with itself under the twist action of the cyclic group of order 2 sending $(x,y)$ to $(y,x)$.

An action of a group $G$ on a locally compact space $X$ is *cocompact* if there exists a compact subset $A$ of $X$ such that $GA = X$. For a properly discontinuous action, cocompactness is equivalent to compactness of the quotient space $X/G$.

The action of $G$ on $X$ is said to be *proper* if the mapping $G \times X \to X \times X$ that sends $(g,x) \mapsto (gx,x)$ is a proper map.

## Strongly continuous group action and smooth points

If is an action of a topological group on another topological space , one says that it is *strongly continuous* if for all , the map $g \mapsto \alpha_g(x)$ is continuous with respect to the respective topologies. Such an action induces an action on the space of continuous function on by .

The subspace of *smooth points* for the action is the subspace of of points such that $g \mapsto \alpha_g(x)$ is smooth; i.e., it is continuous and all derivatives are continuous.

## Generalizations

One can also consider actions of monoids on sets, by using the same two axioms as above. This does not define bijective maps and equivalence relations however. See semigroup action.

Instead of actions on sets, one can define actions of groups and monoids on objects of an arbitrary category: start with an object $X$ of some category, and then define an action on $X$ as a monoid homomorphism into the monoid of endomorphisms of $X$. If $X$ has an underlying set, then all definitions and facts stated above can be carried over. For example, if we take the category of vector spaces, we obtain group representations in this fashion.

One can view a group $G$ as a category with a single object in which every morphism is invertible. A group action is then nothing but a functor from $G$ to the category of sets, and a group representation is a functor from $G$ to the category of vector spaces. A morphism between G-sets is then a natural transformation between the group action functors. In analogy, an action of a groupoid is a functor from the groupoid to the category of sets or to some other category.

Without using the language of categories, one can extend the notion of a group action on a set $X$ by studying as well its induced action on the power set of $X$. This is useful, for instance, in studying the action of the large Mathieu group on a 24-set and in studying symmetry in certain models of finite geometries.

## See also

- Gain graph
- Group with operators
- Monoid action

## Notes

[1]  Eie & Chang (2010), p. 144 (http://books.google.com/books?id=jozIZ0qrkk8C&pg=PA144&dq="group+action")

[2]  This is done e.g. by Smith (2008), p. 253 (http://books.google.com/books?id=PQUAQh04lrUC&pg=PA253&dq="group+action")

[3]  Eie & Chang (2010), p. 145 (http://books.google.com/books?id=jozIZ0qrkk8C&pg=PA144&dq="trivial+action")

## References


- Aschbacher, Michael (2000). *Finite Group Theory*. Cambridge University Press. ISBN 978-0-521-78675-1. MR1777008
- Brown, Ronald (2006). *Topology and groupoids* (http://www.bangor.ac.uk/r.brown/topgpds.html), Booksurge PLC, ISBN 1-4196-2722-8.
- Categories and groupoids, P.J. Higgins (http://138.73.27.39/tac/reprints/articles/7/tr7abs.html), downloadable reprint of van Nostrand Notes in Mathematics, 1971, which deal with applications of groupoids in group theory and topology.
- Dummit, David; Richard Foote (2003). *Abstract Algebra* ((3rd ed.) ed.). Wiley. ISBN 0-471-43334-9.
- Rotman, Joseph (1995). *An Introduction to the Theory of Groups*. Graduate Texts in Mathematics **148** ((4th ed.) ed.). Springer-Verlag. ISBN 0-387-94285-8.
- Smith, Jonathan D.H. (2008). *Introduction to abstract algebra*. Textbooks in mathematics. CRC Press. ISBN 9781420063714.
- Eie, Minking; Chang, Shou-Te (2010). *A Course on Abstract Algebra*. World Scientific. ISBN 9789814271882.
- Weisstein, Eric W., " Group Action (http://mathworld.wolfram.com/GroupAction.html)" from MathWorld.

# Jon Folkman

| Jon Hal Folkman | |
| --- | --- |
| **Born** | December 8, 1938Ogden, Weber County, Utah[1] |
| **Died** | January 23, 1969 (aged 30) |
| **Residence** | United States |
| **Nationality** | ▆▆ United States |
| **Fields** | Combinatorics |
| **Institutions** | RAND Corporation |
| **Alma mater** | Princeton University |
| **Doctoral advisor** | John Milnor |
| **Known for** | Folkman graph<br>Shapley–Folkman lemma & theorem<br>Folkman–Lawrence representation<br>Folkman's theorem (memorial)<br>Homology of lattices and matroids |
| **Notable awards** | Putnam Fellow (1960) |

**Jon Hal Folkman** (December 8, 1938 – January 23, 1969)[2] was an American mathematician, a student of John Milnor, and a researcher at the RAND Corporation.

## Schooling

Folkman was a Putnam Fellow in 1960.[3] He received his Ph.D. in 1964 from Princeton University, under the supervision of Milnor, with a thesis entitled *Equivariant Maps of Spheres into the Classical Groups.*[4]

## Research

Jon Folkman contributed important theorems in many areas of combinatorics.

In geometric combinatorics, Folkman is known for his pioneering and posthumously-published studies of oriented matroids; in particular, the Folkman–Lawrence representation theorem[5] is "one of the cornerstones of the theory of oriented matroids".[6] [7] In lattice theory, Folkman solved an open problem on the foundations of combinatorics by proving a conjecture of Gian–Carlo Rota; in proving Rota's conjecture, Folkman characterized the structure of the homology groups of "geometric lattices" in terms of the free Abelian groups of finite rank.[8] In graph theory, he was the first to study semi-symmetric graphs, and he discovered the semi-symmetric graph with the fewest possible vertices, now known as the Folkman graph.[9] He proved the existence, for every positive $h$, of a finite $K_{h+1}$-free graph which has a monocolored $K_h$ in every 2-coloring of the edges, settling a problem

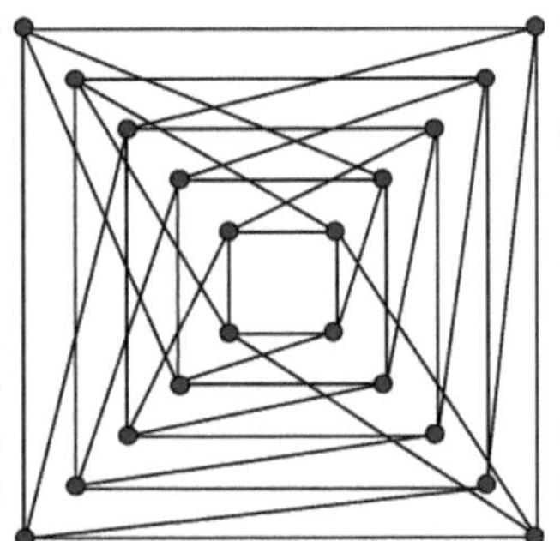

Jon Folkman found the semi-symmetric graph with the fewest possible vertices, the Folkman graph.

previously posed by Paul Erdős and András Hajnal.[10] He further proved that if $G$ is a finite graph such that every set $S$ of vertices contains an independent set of size $(|S| - k)/2$ then the chromatic number of $G$ is at most $k + 2$.[11]

In convex geometry, Folkman worked with his RAND colleague Lloyd Shapley to prove the Shapley–Folkman lemma and theorem: Their results suggest that sums of sets are approximately convex; in mathematical economics their results are used to explain why economies with many agents have approximate equilibria, despite individual nonconvexities.[12]

In additive combinatorics, Folkman's theorem states that for each assignment of finitely many colors to the positive integers, there exist arbitrarily large sets of integers all of whose nonempty sums have the same color; the name was chosen as a memorial to Folkman by his friends.[13] In Ramsey theory, the Rado–Folkman–Sanders theorem describes "partition regular" sets.

## Brain cancer and despair

In the late 1960s, Folkman suffered from brain cancer; while hospitalized, Folkman was visited repeatedly by Ronald Graham and Paul Erdős. After his brain surgery, Folkman was despairing that he had lost his mathematical skills. As soon as Folkman received Graham and Erdős at the hospital, Erdős challenged Folkman with mathematical problems, helping to rebuild his confidence.

Notwithstanding his ability to solve the problems posed by Erdős, Folkman purchased a gun and killed himself. Folkman's supervisor at RAND, Delbert Ray Fulkerson, blamed himself for failing to notice suicidal behaviors in Folkman. Years later Fulkerson also killed himself.[14]

## References

Paul Erdős visited Jon Folkman after Folkman awoke from surgery for brain cancer. To restore Folkman's confidence, Erdős immediately challenged him to solve mathematical problems.[14]


[1]  Jon Hal Folkman (http://www.familysearch.org/Eng/Search/af/ individual_record.asp?recid=5829079&lds=0®ion=-1®ionfriendly=& juris1=&juris2=&juris3=&juris4=®ionfriendly=&juris1friendly=& juris2friendly=&juris3friendly=&juris4friendly=) at *FamilySearch*

[2]  Birth and death dates from Graham, R. L.; Rothschild, B. L., "Ramsey's theorem for *n*-parameter sets" (http://www.math.ucsd.edu/~sbutler/ron/71_04_n_ramsey.pdf), *Transactions of the American Mathematical Society* **159**: 257–292, , and from Spencer, Joel (1971), "Optimal ranking of tournaments", *Networks* **1** (2): 135–138, doi:10.1002/net.3230010204, both of which were dedicated to the memory of Folkman.

[3]  Putnam competition results (http://www.maa.org/awards/putnam.html), Mathematical Association of America, retrieved 2010-10-17.

[4]  John Hal Folkman (http://genealogy.math.ndsu.nodak.edu/id.php?id=22443) at the Mathematics Genealogy Project..

[5]  Folkman, J.; Lawrence, J. (1978), "Oriented matroids", *Journal of Combinatorial Theory, Series B* **25** (2): 199–236, doi:10.1016/0095-8956(78)90039-4.

[6]  Page 17: Björner, Anders; Las Vergnas, Michel; Sturmfels, Bernd; White, Neil; Ziegler, Günter (1999). *Oriented Matroids*. Cambridge University Press. ISBN 9780521777506.

[7]  The Folkman-Lawrence representation theorem is called the "Lawrence representation theorem" by Gunter M. Ziegler in remark 7.23 on page 211: *Lectures on Polytopes*. Graduate texts in mathematics. **152**. New York: Springer-Verlag. 1995. ISBN 0-387-94365-X (paper), 0-387-94329-3.

- Kung, Joseph P. S. (ed.) (1986). "III Enumeration in geometric lattices, 2. Homology". *A Source book in matroid theory*. Boston, MA: Birkhäuser Boston, Inc.. pp. 201–202. ISBN 0-8176-3173-9. MR890330.

  - Folkman, Jon (1966). "The homology groups of a lattice". *Journal of Mathematics and Mechanics* **15**: pp. 631–636. MR188116.
  - Folkman, Jon; Kung, Joseph P. S. (ed.) (1986). "The homology groups of a lattice". *A Source book in matroid theory*. Boston, MA: Birkhäuser Boston, Inc.. pp. 243–248. ISBN 0-8176-3173-9. MR188116.
  - Rota, Gian-Carlo (1964). "On the foundations of combinatorial theory, I: Theory of Möbius functions". *Zeitschrift für Wahrscheinlichkeitstheorie und Verwandte Gebiete (Probability theory and related fields)* **2**: pp. 340–368. doi:10.1007/BF00531932. MR174487.
  - Rota, Gian-Carlo; Kung, Joseph P. S. (ed.) (1986). "On the foundations of combinatorial theory, I: Theory of Möbius functions". *A Source book in matroid theory*. Boston, MA: Birkhäuser Boston, Inc.. pp. 213–242. doi:10.1007/BF00531932. ISBN 0-8176-3173-9.

MR174487.

[9]  Folkman, J. (1967), "Regular line-symmetric graphs", *Journal of Combinatorial Theory* **3**: 215–232, doi:10.1016/S0021-9800(67)80069-3.

[10]  Folkman, J. (1970), "Graphs with monochromatic complete subgraphs in every edge coloring", *SIAM J. Appl. Math.* **18**: 19–24, MR0268080.

[11]  J. Folkman: An upper bound on the chromatic number of a graph, in: Combinatorial theory and its application, II (Proc. Colloq., Balatonfüred, 1969), North-Holland, Amsterdam, 1970, 437–457.

[12]  Starr, Ross M. (1969), "Quasi-equilibria in markets with non-convex preferences (Appendix 2: The Shapley–Folkman theorem, pp. 35–37)", *Econometrica* **37** (1): 25–38, JSTOR 1909201.

[13]  Page 81 in Graham, R.; Rothschild, B.; Spencer, J. H. (1990), *Ramsey Theory* (2nd ed.), New York: John Wiley and Sons, ISBN 0471500461.

[14]  Hoffman, Paul (1998), *The man who loved only numbers: the story of Paul Erdős and the search for mathematical truth*, Hyperion, pp. 109–110, ISBN 9780786863624.

# Folkman graph

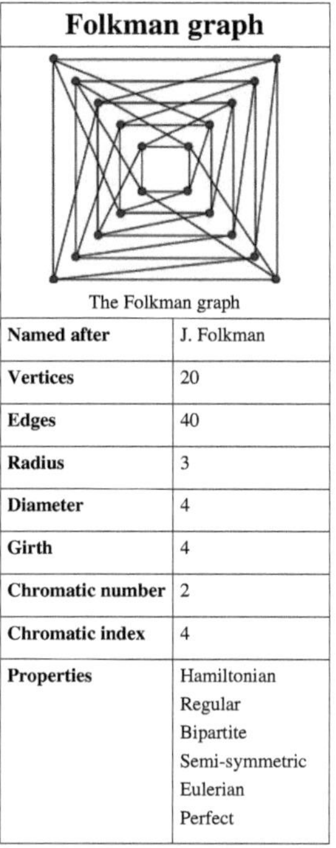

**Folkman graph**

The Folkman graph

| | |
|---|---|
| **Named after** | J. Folkman |
| **Vertices** | 20 |
| **Edges** | 40 |
| **Radius** | 3 |
| **Diameter** | 4 |
| **Girth** | 4 |
| **Chromatic number** | 2 |
| **Chromatic index** | 4 |
| **Properties** | Hamiltonian<br>Regular<br>Bipartite<br>Semi-symmetric<br>Eulerian<br>Perfect |

In the mathematical field of graph theory, the **Folkman graph**, named after Jon Folkman, is a bipartite 4-regular graph with 20 vertices and 40 edges.[1]

The Folkman graph is Hamiltonian and has chromatic number 2, chromatic index 4, radius 3, diameter 4 and girth 4. It is also a 4-vertex-connected and 4-edge-connected perfect graph.

## Algebraic properties

The automorphism group of the Folkman graph graph acts transitively on its edges but not on its vertices. It is the smallest undirected graph that is edge-transitive and regular, but not vertex-transitive.[2] Such graphs are called semi-symmetric graphs and were first studied by Folkman in 1967 who discovered the graph on 20 vertices that now is named after him.[3]

As a semi-symmetric graph, the Folkman graph is bipartite, and its automorphism group acts transitively on each of the two vertex sets of the bipartition. In the diagram below indicating the chromatic number of the graph, the green vertices can not be mapped to red ones by any automorphism, but any red vertex can be mapped on any other red vertex and any green vertex can be mapped on any other green vertex.

The characteristic polynomial of the Folkman graph is .

## Gallery

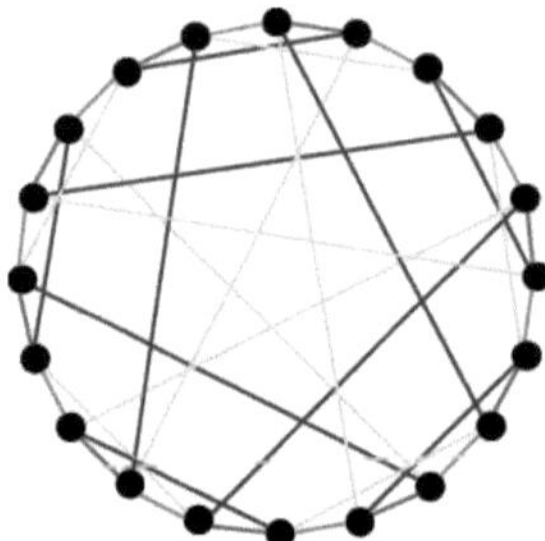

The chromatic index of the Folkman graph is 4.

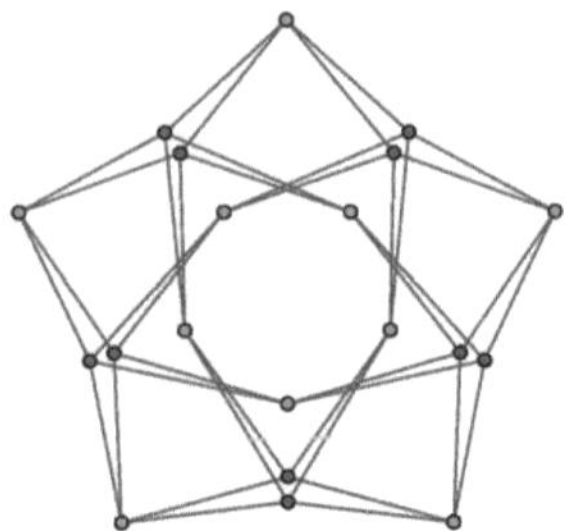

The chromatic number of the Folkman graph is 2.

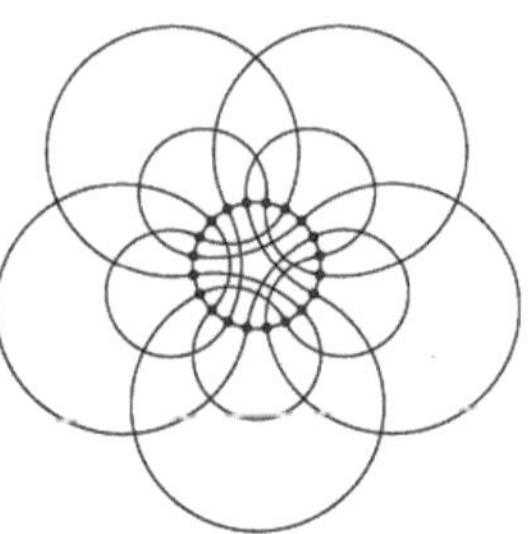

The Folkman graph is Hamiltonian.

## References

[1] Weisstein, Eric W., " Folkman graph (http://mathworld.wolfram.com/FolkmanGraph.html)" from MathWorld.

[2] Skiena, S. Implementing Discrete Mathematics: Combinatorics and Graph Theory with Mathematica. Reading, MA: Addison-Wesley, pp. 186-187, 1990

[3] Folkman, J. (1967), "Regular line-symmetric graphs", *Journal of Combinatorial Theory* **3** (3): 215–232, doi:10.1016/S0021-9800(67)80069-3

# Gray graph

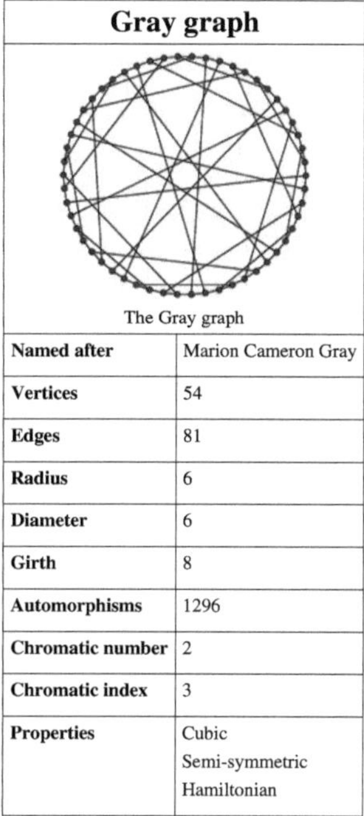

<table>
<tr><td colspan="2" align="center">Gray graph</td></tr>
<tr><td colspan="2" align="center">The Gray graph</td></tr>
<tr><td>Named after</td><td>Marion Cameron Gray</td></tr>
<tr><td>Vertices</td><td>54</td></tr>
<tr><td>Edges</td><td>81</td></tr>
<tr><td>Radius</td><td>6</td></tr>
<tr><td>Diameter</td><td>6</td></tr>
<tr><td>Girth</td><td>8</td></tr>
<tr><td>Automorphisms</td><td>1296</td></tr>
<tr><td>Chromatic number</td><td>2</td></tr>
<tr><td>Chromatic index</td><td>3</td></tr>
<tr><td>Properties</td><td>Cubic<br>Semi-symmetric<br>Hamiltonian</td></tr>
</table>

In the mathematical field of graph theory, the **Gray graph** is an undirected bipartite graph with 54 vertices and 81 edges. It is a cubic graph: every vertex touches exactly three edges. It was discovered by Marion C. Gray in 1932 (unpublished), then discovered independently by Bouwer 1968 in reply to a question posed by Jon Folkman 1967. The Gray graph is interesting as the first known example of a cubic graph having the algebraic property of being edge but not vertex transitive (see below).

The Gray graph has chromatic number 2, chromatic index 3, radius 6 and diameter 6. It is also a 3-vertex-connected and 3-edge-connected non-planar graph.

## Construction

The Gray graph can be constructed (Bouwer 1972) from the 27 points of a 3×3×3 grid and the 27 axis-parallel lines through these points. This collection of points and lines forms a projective configuration: each point has exactly three lines through it, and each line has exactly three points on it. The Gray graph is the Levi graph of this configuration; it has a vertex for every point and every line of the configuration, and an edge for every pair of a point and a line that touch each other. This construction generalizes (Bouwer 1972) to any dimension n ≥ 3, yielding an n-valent Levi graph with algebraic properties similar to those of the Gray graph. In (Monson,Pisanski,Schulte,Ivic-Weiss 2007), the Gray graph appears as a different sort of Levi graph for the edges and triangular faces of a certain locally toroidal abstract regular 4-polytope. It is therefore the first in an infinite family of similarly constructed cubic graphs.

Marušič and Pisanski (2000) give several alternative methods of constructing the Gray graph. As with any bipartite graph, there are no odd-length cycles, and there are also no cycles of four or six vertices, so the girth of the Gray graph is 8. The simplest oriented surface on which the Gray graph can be embedded has genus 7 (Marušič, Pisanski & Wilson 2005).

The Gray graph is Hamiltonian and can be constructed from the LCF notation:

## Algebraic properties

The automorphism group of the Gray graph is a group of order 1296. It acts transitively on the edges the graph but not on its vertices : there are symmetries taking every edge to any other edge, but not taking every vertex to any other vertex. The vertices that correspond to points of the underlying configuration can only be symmetric to other vertices that correspond to points, and the vertices that correspond to lines can only be symmetric to other vertices that correspond to lines. Therefore, the Gray graph is a semi-symmetric graph, the smallest possible cubic semi-symmetric graph.

The characteristic polynomial of the Gray graph is

## Gallery

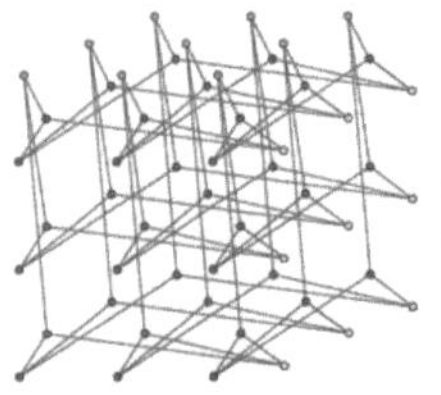

The Gray graph

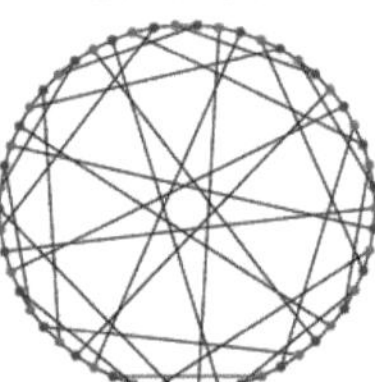

The chromatic number of the Gray graph is 2.

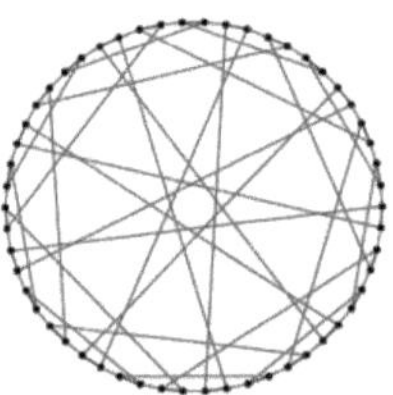

The chromatic index of the Gray graph is 3.

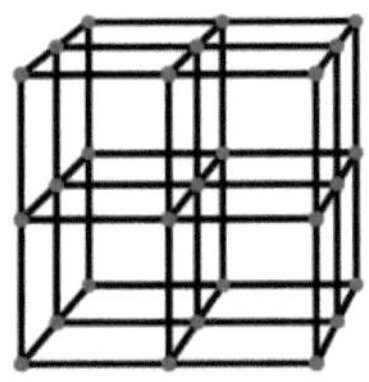

The underlying configuration of the Gray graph.

## References


* Bouwer, I. Z. (1968), "An edge but not vertex transitive cubic graph", *Bulletin of the Canadian Mathematical Society* **11**: 533–535, doi:10.4153/CMB-1968-063-0.
* Bouwer, I. Z. (1972), "On edge but not vertex transitive regular graphs", *Journal of Combinatorial Theory, Series B* **12**: 32–40, doi:10.1016/0095-8956(72)90030-5.
* Folkman, J. (1967), "Regular line-symmetric graphs", *Journal of Combinatorial Theory* **3** (3): 533–535, doi:10.1016/S0021-9800(67)80069-3.
* Marušič, Dragan; Pisanski, Tomaž (2000), "The Gray graph revisited", *Journal of Graph Theory* **35**: 1–7, doi:10.1002/1097-0118(200009)35:1<1::AID-JGT1>3.0.CO;2-7.
* Marušič, Dragan; Pisanski, Tomaž; Wilson, Steve (2005), "The genus of the Gray graph is 7", *European Journal of Combinatorics* **26** (3–4): 377–385, doi:10.1016/j.ejc.2004.01.015.
* Monson, B.; Pisanski, T.; Schulte, E.; Ivic-Weiss, A. (2007), "Semisymmetric Graphs from Polytopes", *Journal of Combinatorial Theory, Series A* **114** (3): 421–435, doi:10.1016/j.jcta.2006.06.007

## External links

- Weisstein, Eric W., "Gray Graph [1]" from MathWorld.
- The Gray Graph Is the Smallest Graph of Its Kind [2], from MathWorld.

## References

[1] http://mathworld.wolfram.com/GrayGraph.html
[2] http://mathworld.wolfram.com/news/2002-04-09/graygraph/

# Dragan Marušič

**Dragan Marušič** (born 1953) is a Slovene mathematician.

His research focuses on topics in algebraic graph theory, particularly the symmetry of graphs and the action of finite groups on combinatorial objects. In 2002, he helped show that the Gray graph is the smallest cubic semi-symmetric graph, resolving a long-open problem.

From 1968 to 1972 Marušič attended gymnasium in Koper. He studied undergraduate mathematics at the University of Ljubljana. He completed his Ph.D. at the University of Reading under the supervision of Crispin Nash-Williams.

Marušič is regarded as the founder of the Slovenian school of research in algebraic graph theory and permutation groups. Currently, he is on the faculty of the University of Ljubljana.

In 2002 he received the Zois award for his achievements in the field of graph theory and algebra; founding editor of the *Ars Mathematica Contemporanea*.

## External links

- Dragan Marušič [1] at the Mathematics Genealogy Project.
- personal web page [2]

## References

[1] http://genealogy.math.ndsu.nodak.edu/id.php?id=37105
[2] http://draganmarusic.si

- Malnič, A.; Marušič, D.; Potočnik, P.; and Wang, C. (2002). "An Infinite Family of Cubic Edge-Transitive but Not Vertex-Transitive Graphs". *Discr. Math.*.

# Ljubljana graph

### Ljubljana graph

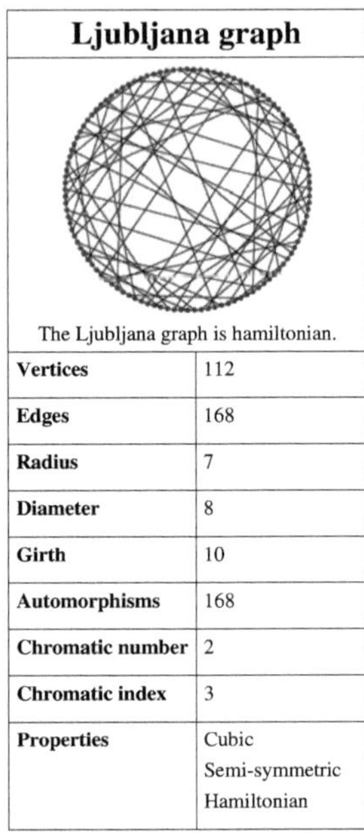

The Ljubljana graph is hamiltonian.

| | |
|---|---|
| **Vertices** | 112 |
| **Edges** | 168 |
| **Radius** | 7 |
| **Diameter** | 8 |
| **Girth** | 10 |
| **Automorphisms** | 168 |
| **Chromatic number** | 2 |
| **Chromatic index** | 3 |
| **Properties** | Cubic<br>Semi-symmetric<br>Hamiltonian |

In the mathematical field of graph theory, the **Ljubljana graph** is an undirected bipartite graph with 112 vertices and 168 edges.[1]

It is a cubic graph with diameter 8, radius 7, chromatic number 2 and chromatic index 3. Its girth is 10 and there are exactly 168 cycles of length 10 in it. There are also 168 cycles of length 12.[2]

## Construction

The Ljubljana graph is Hamiltonian and can be constructed from the LCF notation : [47, -23, -31, 39, 25, -21, -31, -41, 25, 15, 29, -41, -19, 15, -49, 33, 39, -35, -21, 17, -33, 49, 41, 31, -15, -29, 41, 31, -15, -25, 21, 31, -51, -25, 23, 9, -17, 51, 35, -29, 21, -51, -39, 33, -9, -51, 51, -47, -33, 19, 51, -21, 29, 21, -31, -39]$^2$.

The Ljubljana graph is the Levi graph of the Ljubljana configuration, a quadrangle-free configuration with 56 lines and 56 points.[2] In this configuration, each line contains exactly 3 points, each points belongs to exactly 3 lines and any two lines intersect in at most one point.

## Algebraic properties

The automorphism group of the Ljubljana graph is a group of order 168. It acts transitively on the edges the graph but not on its vertices : there are symmetries taking every edge to any other edge, but not taking every vertex to any other vertex. Therefore, the Ljubljana graph is a semi-symmetric graph, the third smallest possible cubic semi-symmetric graph after the Gray graph on 54 vertices and the Iofinova-Ivanov graph on 110 vertices.[3]

The characteristic polynomial of the Ljubljana graph is

## History

The Ljubljana graph was first published in 1993 by Brouwer, Dejter and Thomassen.[4]

In 1972, Bouwer was already talking of a 112-vertices edge- but not vertex-transitive cubic graph found by R. M. Foster, but unpublished.[5] Conder, Malnič, Marušič, Pisanski and Potočnik rediscovered this 112-vertices graph in 2002 and named it the Ljubljana graph after the capital of Slovenia.[2] They proved that it was the unique 112-vertices edge- but not vertex-transitive cubic graph and therefore that was the graph found by Foster.

## Gallery

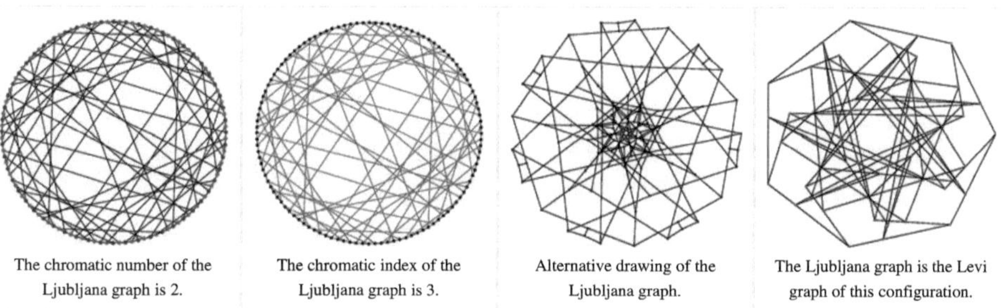

| The chromatic number of the Ljubljana graph is 2. | The chromatic index of the Ljubljana graph is 3. | Alternative drawing of the Ljubljana graph. | The Ljubljana graph is the Levi graph of this configuration. |

## References

[1]  Weisstein, Eric W., " Ljubljana Graph (http://mathworld.wolfram.com/LjubljanaGraph.html)" from MathWorld.

[2]  Conder, M.; Malnič, A.; Marušič, D.; Pisanski, T.; and Potočnik, P. "The Ljubljana Graph." 2002. (http://citeseer.ist.psu.edu/conder02ljubljana.html).

[3]  Marston Conder, Aleksander Malnič, Dragan Marušič and Primž Potočnik. "A census of semisymmetric cubic graphs on up to 768 vertices." Journal of Algebraic Combinatorics: An International Journal. Volume 23, Issue 3, pages 255-294, 2006.

[4]  Brouwer, A. E.; Dejter, I. J.; and Thomassen, C. "Highly Symmetric Subgraphs of Hypercubes." J. Algebraic Combinat. 2, 25-29, 1993.

[5]  Bouwer, I. A. "On Edge But Not Vertex Transitive Regular Graphs." J. Combin. Th. Ser. B 12, 32-40, 1972.

# Tutte 12-cage

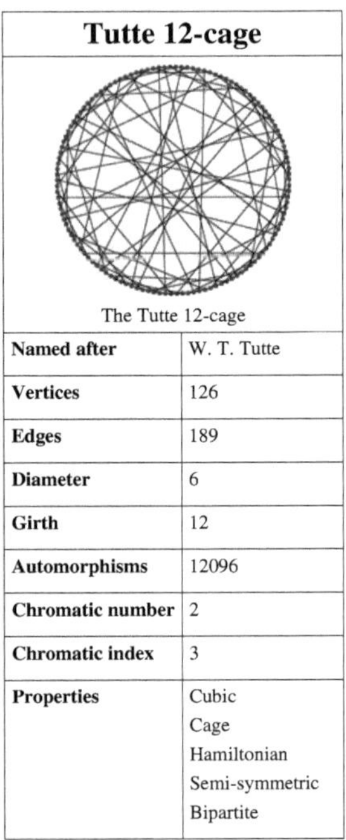

|  |  |
|---|---|
| **Tutte 12-cage** | |
| The Tutte 12-cage | |
| **Named after** | W. T. Tutte |
| **Vertices** | 126 |
| **Edges** | 189 |
| **Diameter** | 6 |
| **Girth** | 12 |
| **Automorphisms** | 12096 |
| **Chromatic number** | 2 |
| **Chromatic index** | 3 |
| **Properties** | Cubic<br>Cage<br>Hamiltonian<br>Semi-symmetric<br>Bipartite |

In the mathematical field of graph theory, the **Tutte 12-cage** or **Benson graph**[1] is a 3-regular graph with 126 vertices and 189 edges named after W. T. Tutte.[2]

The Tutte 12-cage is the unique (3-12)-cage (sequence A052453 in OEIS). It was discovered by C. T. Benson in 1966.[3] It has chromatic number 2 (bipartite), chromatic index 3, girth 12 (as a 12-cage) and diameter 6. Its crossing number is 170 and has been conjectured to be the smallest cubic graph with this crossing number.[4] [5]

## Construction

The Tutte 12-cage is a cubic Hamiltonian graph and can be defined by the LCF notation [17, 27, −13, −59, −35, 35, −11, 13, −53, 53, −27, 21, 57, 11, −21, −57, 59, −17]$^7$.[6]

It is the incidence graph of the generalized hexagon GH(2,2) with 63 points and 63 lines.[1]

The Balaban 11-cage can be constructed by excision from the Tutte 12-cage by removing a small subtree and suppressing the resulting vertices of degree two.[7]

## Algebraic properties

The automorphism group of the Tutte 12-cage is of order 12096 and is a semi-direct product of the projective special unitary group PSU(3,3) with the cyclic group **Z**/2**Z**.[1] Its acts transitively on its edges but not on its vertices, making it a semi-symmetric graph, a regular graph that is edge-transitive but not vertex-transitive. In fact, the automorphism group of the Tutte 12-cage preserves the bipartite parts and acts primitively on each part. Such graphs are called bi-primitive graphs and only five cubic bi-primitive graphs exist; they are named the Iofinova-Ivanov graphs and are of order 110, 126, 182, 506 and 990.[8]

All the cubic semi-symmetric graphs on up to 768 vertices are known. According to Conder, Malnič, Marušič and Potočnik, the Tutte 12-cage is the unique cubic semi-symmetric graph on 126 vertices and is the fifth smallest possible cubic semi-symmetric graph after the Gray graph, the Iofinova–Ivanov graph on 110 vertices, the Ljubljana graph and a graph on 120 vertices with girth 8.[9]

The characteristic polynomial of the Tutte 12-cage is

It is the only graph with this characteristic polynomial; therefore, the 12-cage is determined by its spectrum.

## Gallery

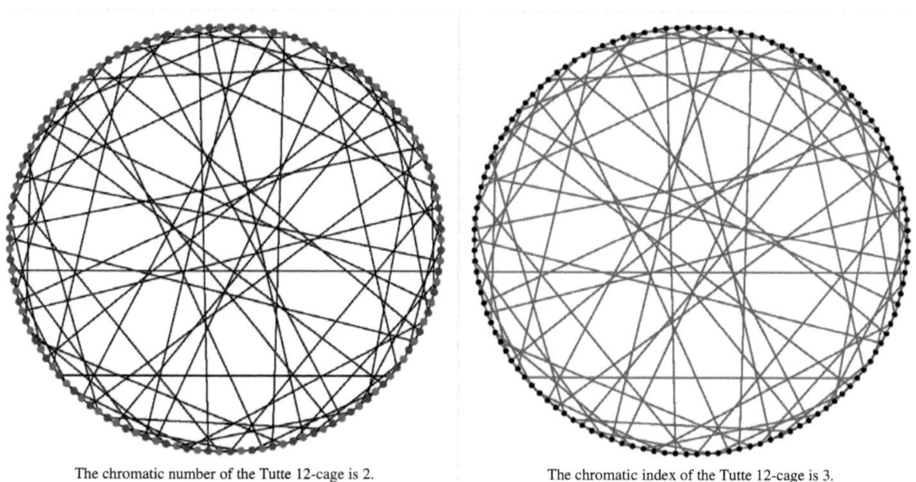

The chromatic number of the Tutte 12-cage is 2.

The chromatic index of the Tutte 12-cage is 3.

## References

[1]  Geoffrey Exoo & Robert Jajcay, Dynamic cage survey, Electr. J. Combin. 15 (2008).

[2]  Weisstein, Eric W., " Tutte 12-cage (http://mathworld.wolfram.com/Tutte12-Cage.html)" from MathWorld.

[3]  Benson, C. T. "Minimal Regular Graphs of Girth 8 and 12." Canad. J. Math. 18, 1091–1094, 1966.

[4]  Exoo, G. "Rectilinear Drawings of Famous Graphs" (http://isu.indstate.edu/ge/COMBIN/RECTILINEAR/).

[5]  Pegg, E. T. and Exoo, G. "Crossing Number Graphs." Mathematica J. 11, 2009.

[6]  Polster, B. A Geometrical Picture Book. New York: Springer, p. 179, 1998.

[7]  Balaban, A. T. "Trivalent Graphs of Girth Nine and Eleven and Relationships Among the Cages." Rev. Roumaine Math 18, 1033–1043, 1973.

[8]  Iofinova, M. E. and Ivanov, A. A. "Bi-Primitive Cubic Graphs." In Investigations in the Algebraic Theory of Combinatorial Objects. pp. 123–134, 2002. (Vsesoyuz. Nauchno-Issled. Inst. Sistem. Issled., Moscow, pp. 137–152, 1985.)

[9]  Conder, Marston; Malnič, Aleksander; Marušič, Dragan; Potočnik, Primož (2006), "A census of semisymmetric cubic graphs on up to 768 vertices", *Journal of Algebraic Combinatorics* **23**: 255–294, doi:10.1007/s10801-006-7397-3.

# Article Sources and Contributors

**Semi-symmetric graph** *Source*: http://en.wikipedia.org/w/index.php?title=Semi-symmetric_graph *Contributors*: David Eppstein, Koko90, Michael Hardy, Miym, Radagast3, Ryan Reich, Silverfish, Tomo, Twri, 3 anonymous edits

**Graph theory** *Source*: http://en.wikipedia.org/w/index.php?title=Graph_theory *Contributors*: APH, Aaronzat, Abeg92, Agro1986, Ajweinstein, Aknxy, Akutagawa10, Alan Au, Alansohn, Alcidesfonseca, Alex Bakharev, Almi, Altenmann, Ams80, Andre Engels, Andrea105, Andreas Kaufmann, Andris, Ankitbhatt, Anna Lincoln, Anonymous Dissident, Anthonynow12, Aquae, Arbor, Armoreno10, Arvindn, Astrophil, AxelBoldt, Ayda D, Bact, Beda42, Bender235, Bereziny, Berteun, Bkell, Blair Sutton, BlckKnght, Boleslav Bobcik, Booyabazooka, Brad7777, Brent Gulanowski, Brick Thrower, Brona, Bumbulski, C S, CRGreathouse, Camembert, CanadianLinuxUser, Caramdir, Cerber, CesarB, Chalst, Charles Matthews, Chas zzz brown, Chmod007, Conversion script, Corpx, Csl77, D75304, DFRussia, Damien Karras, Daniel Quinlan, David Eppstein, Davidfstr, Dbenbenn, Delaszk, DerHexer, Dgrant, Dicklyon, Diego UFCG, Dina, Disavian, Discospinster, Dittaeva, Doctorbozzball, Dr.S.Ramachandran, Duncharris, Dycedarg, Dysprosia, Dze27, ElBenevolente, Eleuther, Elf, Eric.weigle, Eugeneiiim, Favonian, FayssalF, Fchristo, Feeeshboy, Fivelittlemonkeys, Fred Bradstadt, Fredrik, FvdP, GGordonWorleyIII, GTBacchus, Gaius Cornelius, Gandalf61, Garyzx, Geometry guy, George Burgess, Gianfranco, Giftlite, GiveAFishABone, Glinos, Gmelli, Goochelaar, Gragragra, Graham87, GraphTheoryPwns, GregorB, Grog, Gutza, Gyan, Hannes Eder, Hans Adler, Hazmat2, Hbruhn, Headbomb, Henning Makholm, Hirzel, Idiosyncratic-bumblebee, Ignatzmice, Igodard, Ino5hiro, Jakob Voss, Jalpar75, Jaxl, JeLuF, Jeronimo, Jheuristic, Jinma, Joel B. Lewis, Joespiff, JohnBlackburne, Johnsopc, Jojit fb, Jon Awbrey, Jon har, Jorge Stolfi, Jwissick, Kalogeropoulos, Karl E. V. Palmen, KellyCoinGuy, King Bee, Knutux, Kope, Kpjas, Kruusamägi, LC, LOL, Lanem, Ldecola, Liao, Linas, Looxix, Low-frequency internal, Lpgeffen, Lupin, Madewokherd, Magister Mathematicae, Magmi, Mahlerite, Mailer diablo, Mani1, Marek69, Marianocecowski, Mark Foskey, Mark Renier, MathMartin, Matusz, Maurobio, Maxime.Debosschere, McKay, Meekohi, Melchoir, Michael Hardy, Michael Slone, Miguel, Mild Bill Hiccup, Mlym, Mlpkr, Morphh, Msh210, Mxn, Myanw, MynameisJayden, Naerb14c, Nanshu, Nasnema, Nichtich, Ntsimp, Obankston, Obradovic Goran, Ohnoitsjamie, Oleg Alexandrov, Oli Filth, Onorem, Optim, OrangeDog, Oroso, Oskar Flordal, Outraged duck, Pashute, Patrick, Paul August, Paul Murray, PaulHoadley, PaulTanenbaum, Pcb21, Peter Kwok, Piano non troppo, PierreAbbat, Poor Yorick, Populus, Powerthirst123, Protonk, Pstanton, Puckly, Pugget, Quaeler, Qutezuce, R'n'B, RUVARD, Radagast3, RandomAct, Ratiocinate, Renice, Requestion, Restname, Rev3nant, RexNL, Rjwilmsi, Rmiesen, Roachmeister, Robert Merkel, Robin klein, RobinK, Ronz, Ruud Koot, SCEhardt, Sacredmint, Sangak, Shanes, Shd, Shepazu, Shikhar1986, Shizhao, Sibian, Slawekb, SlumdogAramis, Smoke73, Solitude, Sonia, Spacefarer, Stochata, Sundar, Sverdrup, TakuyaMurata, Tangi-tamma, Tarotcards, Taw, Taxipom, Tckma, Template namespace initialisation script, The Cave Troll, The Isiah, Thesilverbail, Thv, Titanic4000, Tomo, Tompw, Trinitrix, Tyir, Tyler McHenry, Uncle Dick, Usien6, Vonkje, Watersmeetfreak, Whiteknox, Whyfish, Womiller99, Wshun, Wsu-dm-jb, Wsu-f, XJamRastafire, Xdenizen, Xiong, XxjwuxX, Yecril, Ylloh, Youngster68, Zaslav, Zero0000, Zoicon5, Zundark, Александър, Канеюку, 429 anonymous edits

**Graph (mathematics)** *Source*: http://en.wikipedia.org/w/index.php?title=Graph_%28mathematics%29 *Contributors*: 3ICE, ABF, Abdull, Ahoerstemeier, Aitias, Ajraddatz, Aknorals, Akutagawa10, Algont, Altenmann, Amintora, Anabus, Anand jeyahar, Andrewsky00, Andros 1337, Athenray, Barras, BenRG, Bethnim, Bhadani, BiT, Bkell, Bobo192, Booyabazooka, Borgx, Brentdax, Burnin1134, CRGreathouse, Can't sleep, clown will eat me, Catgut, Cbdorsett, Cbuckley, Ch'marr, Chinasaur, Chmod007, Chris the speller, Chronist, Citrus538, Cjfsyntropy, Cornflake pirate, Corti, Crisófilax, Cybercobra, DARTH SIDIOUS 2, Danrah, David Eppstein, Davidfstr, Dbenbenn, Dcoetzee, Ddxc, Debamf, Debeolaurus, Den fjättrade ankan, Deor, Dicklyon, Djk3, Dockfish, Dreamster, Dtrebbien, Dureo, Dysprosia, E mraedarab, Editor70, Erhudy, Eric Lengyel, Falcon8765, Gaius Cornelius, Gandalf61, Gauge, Gene.arboit, George100, Giftlite, Grolmusz, Gutin, Gwaihir, Hairy Dude, Hannes Eder, Hans Adler, Hans Dunkelberg, Harp, Headbomb, Henry Delforn, Huynl, Ijdejter, Ilia Kr., Ilya, Insanity Incarnate, J.delanoy, JNW, Jiang, Joeblakesley, Jokes Free4Me, Jon Awbrey, Jon har, Joriki, Jpeeling, Jpiw, JuPitEer, Jwanders, Karada, Karlscherer3, Kine, Kingmash, Knutux, Kruusamägi, Kuru, L Kensington, Liao, Libcub, Lipedia, LuckyWizard, MER-C, Maghnus, Magister Mathematicae, Mahanga, Marc van Leeuwen, Materialscientist, MathMartin, Matt Crypto, McKay, Mdd, Michael Hardy, Michael Slone, MikeBorkowski, Minder2k, Mindmatrix, Mitmaro, Miym, Mountain, Myasuda, Mymyhoward16, Neilc, Netrapt, Nowhither, Nsk92, Ohnoitsjamie, Oliphaunt, Oxymoron83, Pafcu, Paleorthid, Patrick, Paul August, PaulTanenbaum, Peter Kwok, Peterl, Phegyi81, PhotoBox, Possum, Powerthirst123, Prunesqualer, Quaeler, R'n'B, Radagast3, RaseaC, Repied, Requestion, Rettetast, Rgclegg, Rich Farmbrough, Rjwilmsi, RobertBorgersen, Robertsteadman, RobinK, Royerloic, Salix alba, Sdrucker, Shadowjams, Shanel, Siddhant, Silversmith, Someguy1221, SophomoricPedant, Sswn, Stevertigo, Struthious Bandersnatch, Stux, Suchap, Super-Magician, Svick, THEN WHO WAS PHONE?, Tangi-tamma, Tbc2, Tempodivalse, Tgv8925, The Anome, Theone256, TimBentley, Tman159, Tobias Bergemann, Tom harrison, Tomhubbard, Tomo, Tompw, Tomruen, Tosha, Tslocum, Twri, Tyw7, Ulric1313, Urdutext, Vaughan Pratt, Void-995, W, Wandrer2, Wesley Moy, West.andrew.g, Wgunther, White Trillium, WikiDao, Wikidsp, Willking1979, Wknight94, Wshun, XJamRastafire, Xavexgoem, Xiong, Yath, Yecril, Yitzhak, Ylloh, Zachlipton, Zaslav, Zero0000, Zocky, Zven, Канеюку, Пика Пика, 372 anonymous edits

**Edge-transitive graph** *Source*: http://en.wikipedia.org/w/index.php?title=Edge-transitive_graph *Contributors*: Csab, David Eppstein, Genusfour, Giftlite, Hairy Dude, Jaredwf, Michael Hardy, RJFJR, Radagast3, Silverfish, Steelpillow, Tomo, 5 anonymous edits

**Vertex-transitive graph** *Source*: http://en.wikipedia.org/w/index.php?title=Vertex-transitive_graph *Contributors*: Algebraist, Charles Matthews, Csab, David Eppstein, FRR, Giftlite, Headbomb, Jaredwf, Koko90, Maksim-e, McKay, Mhym, Ntsimp, Phil Boswell, Radagast3, Robert Illes, RobinK, Silverfish, Tomo, Twri, 5 anonymous edits

**Graph automorphism** *Source*: http://en.wikipedia.org/w/index.php?title=Graph_automorphism *Contributors*: Altenmann, David Eppstein, DeC, Giftlite, Hans Adler, Igor Markov, Joel B. Lewis, Joriki, Koko90, LOL, Maudgalya, Mhym, Miym, Ott2, Radagast3, Robert Illes, Twri, Канеюку, 4 anonymous edits

**Group action** *Source*: http://en.wikipedia.org/w/index.php?title=Group_action *Contributors*: 3mta3, Acepectif, Ajstern, Archelon, ArnoldReinhold, AxelBoldt, BenFrantzDale, BenKovitz, Brad7777, Bryan Derksen, CBM, Calle, Cgibbard, Charles Matthews, Chas zzz brown, Chinju, ChrisHodgesUK, Compsonheir, Conversion script, Cullinane, Cwkmail, Dominicanpapi82, Dominus, Dysprosia, ELLinng, Efio, El C, ElNuevoEinstein, Elroch, Esoth, FarSide, Fropuff, Ft1, Gaius Cornelius, Geometry guy, Giansira, Giftlite, Grubber, Haham hanuka, Helder.wiki, Ht686rg90, JLeander, JackSchmidt, Jfr26, Johanbosman, John Baez, Jokes Free4Me, Jtwdog, Kaiwhakahere, Kallikanzarid, Kjetil1001, LC, Lethe, Linas, Loodog, Lotje, MFH, MSGJ, MarSch, Marc van Leeuwen, Marcowongshuinam, Masnevets, MathMartin, Mathchem271828, Mathemajor, Mets501, Michael Hardy, Michael Larsen, MichaelBeck, Mosher, Nbarth, Nick, Nm420, Oleg Alexandrov, Orthografer, Patrick, Pedro Lopez-Adeva Fernandez-Layos, Phys, Piotrus, Piyush Sriva, Pyrop, Rigadoun, RobHar, RonnieBrown, Rs2, Rvollmert, Salix alba, Schutz, Silly rabbit, Simplifix, Sophoric, Spike Wilbury, Sławomir Biały, Taemyr, The Anome, Thehotelambush, Tijfo098, Tobias Bergemann, Toby Bartels, Tong, Tosha, Ultramarine, Vectornaut, Wikfr, Wikiklrsc, Yecril, Zaslav, Zfeinst, Zundark, 99 anonymous edits

**Jon Folkman** *Source*: http://en.wikipedia.org/w/index.php?title=Jon_Folkman *Contributors*: David Eppstein, Headbomb, Kiefer.Wolfowitz, Kope, Maurice Carbonaro, Michael Hardy, Omnipaedista, The Anome, WhatamIdoing, 1 anonymous edits

**Folkman graph** *Source*: http://en.wikipedia.org/w/index.php?title=Folkman_graph *Contributors*: David Eppstein, Giftlite, Kiefer.Wolfowitz, Koko90, Kope, Michael Hardy, The Anome, Twri, XJamRastafire

**Gray graph** *Source*: http://en.wikipedia.org/w/index.php?title=Gray_graph *Contributors*: AlbertR, Booyabazooka, Charles Matthews, David Eppstein, GTBacchus, Giftlite, Joel7687, Koko90, Michael Hardy, Poulpy, Silverfish, Tomo, Twri, 10 anonymous edits

**Dragan Marušič** *Source*: http://en.wikipedia.org/w/index.php?title=Dragan_Maru%C5%A1i%C4%8D *Contributors*: Altenmann, Charles Matthews, D6, Dante Alighieri, David Eppstein, Gadfium, Gooddays, Hillman, Icairns, JYolkowski, Jaredwf, Jengod, Jpbowen, Kbdank71, Naive cynic, Ntsimp, Oleg Alexandrov, Paul August, Remy B, Rich Farmbrough, Saga City, Splash, TheParanoidOne, Tomo, TonyW, Valentinian, XJamRastafire, 5 anonymous edits

**Ljubljana graph** *Source*: http://en.wikipedia.org/w/index.php?title=Ljubljana_graph *Contributors*: Dsp13, Giftlite, Koko90, LilHelpa, Radagast3, Twri, XJamRastafire, 1 anonymous edits

**Tutte 12-cage** *Source*: http://en.wikipedia.org/w/index.php?title=Tutte_12-cage *Contributors*: Giftlite, Koko90, M.nelson, Michael Hardy, Piledhigheranddeeper, Twri

# Image Sources, Licenses and Contributors

**Image:Folkman graph.svg** *Source*: http://en.wikipedia.org/w/index.php?title=File:Folkman_graph.svg *License*: unknown *Contributors*: User:David Eppstein

**Image:6n-graf.svg** *Source*: http://en.wikipedia.org/w/index.php?title=File:6n-graf.svg *License*: unknown *Contributors*: User:AzaToth

**Image:Konigsberg bridges.png** *Source*: http://en.wikipedia.org/w/index.php?title=File:Konigsberg_bridges.png *License*: unknown *Contributors*: Bogdan Giuşcă

**Image:Multigraph.svg** *Source*: http://en.wikipedia.org/w/index.php?title=File:Multigraph.svg *License*: unknown *Contributors*: User:Arthena

**Image:Undirected.svg** *Source*: http://en.wikipedia.org/w/index.php?title=File:Undirected.svg *License*: unknown *Contributors*: JMCC1, Josette, Kilom691, 2 anonymous edits

**Image:Directed.svg** *Source*: http://en.wikipedia.org/w/index.php?title=File:Directed.svg *License*: unknown *Contributors*: Grafite, Jcb, Josette, 2 anonymous edits

**File:Complete graph K5.svg** *Source*: http://en.wikipedia.org/w/index.php?title=File:Complete_graph_K5.svg *License*: unknown *Contributors*: User:Dbenbenn

**Image:Gray graph 2COL.svg** *Source*: http://en.wikipedia.org/w/index.php?title=File:Gray_graph_2COL.svg *License*: unknown *Contributors*: User:Koko90

**Image:TruncatedTetrahedron.gif** *Source*: http://en.wikipedia.org/w/index.php?title=File:TruncatedTetrahedron.gif *License*: unknown *Contributors*: User:Radagast3

**Image:Petersen1_tiny.svg** *Source*: http://en.wikipedia.org/w/index.php?title=File:Petersen1_tiny.svg *License*: unknown *Contributors*: User:Leshabirukov

**File:Dodecahedral graph.neato.svg** *Source*: http://en.wikipedia.org/w/index.php?title=File:Dodecahedral_graph.neato.svg *License*: unknown *Contributors*: User:Koko90

**File:Arrow east.svg** *Source*: http://en.wikipedia.org/w/index.php?title=File:Arrow_east.svg *License*: unknown *Contributors*: User:MarianSigler

**File:Shrikhande graph square.svg** *Source*: http://en.wikipedia.org/w/index.php?title=File:Shrikhande_graph_square.svg *License*: unknown *Contributors*: User:Koko90

**File:Arrow west.svg** *Source*: http://en.wikipedia.org/w/index.php?title=File:Arrow_west.svg *License*: unknown *Contributors*: User:MarianSigler

**File:Paley13 no label.svg** *Source*: http://en.wikipedia.org/w/index.php?title=File:Paley13_no_label.svg *License*: unknown *Contributors*: User:Koko90

**File:Arrow south.svg** *Source*: http://en.wikipedia.org/w/index.php?title=File:Arrow_south.svg *License*: unknown *Contributors*: User:MarianSigler

**File:F26A graph.svg** *Source*: http://en.wikipedia.org/w/index.php?title=File:F26A_graph.svg *License*: unknown *Contributors*: User:Koko90

**File:Nauru graph.svg** *Source*: http://en.wikipedia.org/w/index.php?title=File:Nauru_graph.svg *License*: unknown *Contributors*: User:Koko90

**File:Holt graph.svg** *Source*: http://en.wikipedia.org/w/index.php?title=File:Holt_graph.svg *License*: unknown *Contributors*: User:Koko90

**File:Folkman Lombardi.svg** *Source*: http://en.wikipedia.org/w/index.php?title=File:Folkman_Lombardi.svg *License*: unknown *Contributors*: User:David Eppstein

**File:Biclique_K_3_5.svg** *Source*: http://en.wikipedia.org/w/index.php?title=File:Biclique_K_3_5.svg *License*: unknown *Contributors*: User:Koko90

**File:Truncated tetrahedral graph.circo.svg** *Source*: http://en.wikipedia.org/w/index.php?title=File:Truncated_tetrahedral_graph.circo.svg *License*: unknown *Contributors*: User:Koko90

**File:Frucht graph.neato.svg** *Source*: http://en.wikipedia.org/w/index.php?title=File:Frucht_graph.neato.svg *License*: unknown *Contributors*: User:Koko90

**File:Arrow north.svg** *Source*: http://en.wikipedia.org/w/index.php?title=File:Arrow_north.svg *License*: unknown *Contributors*: User:MarianSigler

**Image:Z 2xZ 3.svg** *Source*: http://en.wikipedia.org/w/index.php?title=File:Z_2xZ_3.svg *License*: unknown *Contributors*: User:Tosha

**File:Group action on equilateral triangle.svg** *Source*: http://en.wikipedia.org/w/index.php?title=File:Group_action_on_equilateral_triangle.svg *License*: unknown *Contributors*: User:Oleg Alexandrov

**File:Compound of five tetrahedra.png** *Source*: http://en.wikipedia.org/w/index.php?title=File:Compound_of_five_tetrahedra.png *License*: unknown *Contributors*: Magog the Ogre, Tomruen, 2 anonymous edits

**File:Flag of the United States.svg** *Source*: http://en.wikipedia.org/w/index.php?title=File:Flag_of_the_United_States.svg *License*: unknown *Contributors*: Anomie

**Image:Folkman graph alt.svg** *Source*: http://en.wikipedia.org/w/index.php?title=File:Folkman_graph_alt.svg *License*: unknown *Contributors*: User:Koko90

**Image:Erdos head budapest fall 1992.jpg** *Source*: http://en.wikipedia.org/w/index.php?title=File:Erdos_head_budapest_fall_1992.jpg *License*: unknown *Contributors*: Kmhkmh

**Image:Folkman graph 4color edge.svg** *Source*: http://en.wikipedia.org/w/index.php?title=File:Folkman_graph_4color_edge.svg *License*: unknown *Contributors*: User:Koko90

**Image:Gray graph hamiltonian.svg** *Source*: http://en.wikipedia.org/w/index.php?title=File:Gray_graph_hamiltonian.svg *License*: unknown *Contributors*: User:Koko90

**File:Gray graph.svg** *Source*: http://en.wikipedia.org/w/index.php?title=File:Gray_graph.svg *License*: unknown *Contributors*: User:David Eppstein

**File:gray_graph_2COL.svg** *Source*: http://en.wikipedia.org/w/index.php?title=File:Gray_graph_2COL.svg *License*: unknown *Contributors*: User:Koko90

**File:gray_graph_3color_edge.svg** *Source*: http://en.wikipedia.org/w/index.php?title=File:Gray_graph_3color_edge.svg *License*: unknown *Contributors*: User:Koko90

**File:Gray configuration.svg** *Source*: http://en.wikipedia.org/w/index.php?title=File:Gray_configuration.svg *License*: unknown *Contributors*: User:David Eppstein

**Image:Ljubljana graph hamiltonian.svg** *Source*: http://en.wikipedia.org/w/index.php?title=File:Ljubljana_graph_hamiltonian.svg *License*: unknown *Contributors*: User:Koko90

**Image:Ljubljana_graph_2COL.svg** *Source*: http://en.wikipedia.org/w/index.php?title=File:Ljubljana_graph_2COL.svg *License*: unknown *Contributors*: User:Koko90

**Image:Ljubljana_graph_3color_edge.svg** *Source*: http://en.wikipedia.org/w/index.php?title=File:Ljubljana_graph_3color_edge.svg *License*: unknown *Contributors*: User:Koko90

**Image:Ljubljana graph.svg** *Source*: http://en.wikipedia.org/w/index.php?title=File:Ljubljana_graph.svg *License*: unknown *Contributors*: User:Koko90

**Image:Ljubljana configuration.svg** *Source*: http://en.wikipedia.org/w/index.php?title=File:Ljubljana_configuration.svg *License*: unknown *Contributors*: User:Koko90

**Image:Tutte 12-cage.svg** *Source*: http://en.wikipedia.org/w/index.php?title=File:Tutte_12-cage.svg *License*: unknown *Contributors*: User:Koko90

**Image:Tutte 12-cage 2COL.svg** *Source*: http://en.wikipedia.org/w/index.php?title=File:Tutte_12-cage_2COL.svg *License*: unknown *Contributors*: User:Koko90

**Image:Tutte 12-cage 3color edge.svg** *Source*: http://en.wikipedia.org/w/index.php?title=File:Tutte_12-cage_3color_edge.svg *License*: unknown *Contributors*: User:Koko90

Printed by Books on Demand GmbH, Norderstedt / Germany